COMMISSION OF THE EUROPEAN COMMUNITIES
Directorate-General for Research, Science and Education

ENERGY RESEARCH AND DEVELOPMENT PROGRAMME

FIRST STATUS REPORT
(1975-1976)

- Energy Conservation
- Production and Utilization of Hydrogen
- Solar Energy
- Geothermal Energy
- Systems Analysis: Development of Models

July 1977

XII/697/77-E

MARTINUS NIJHOFF - THE HAGUE

Compiled by

Commission of the European Communities,
Directorate-General for Research, Science and Education,
Brussels.

Publication arranged by

Commission of the European Communities,
Directorate-General for Scientific and Technical Information and Information Management,
Luxembourg.

EUR 5889
ISBN-13: 978-90-247-2059-0   e-ISBN-13: 978-94-009-9702-8
DOI: 10.1007/978-94-009-9702-8

Legal notice

Preface

The energy R & D programme of the European Communities,
a four year indirect action programme coping with "new
energies", energy conservation and energy systems
modelling, is now well under way. All projects to be
carried out in the frame of the first phase of this
programme are in progress and the Commission's services,
together with the different advisory bodies, are already
strongly involved in the implementation of the second
phase of the programme.

It seemed to us at this point that the publication of a
comprehensive overview of all contracts concluded in the
first programme phase would be the best way of keeping
informed those individuals or institutions naturally
interested in our work. The present report should be
considered as the first attempt to fulfill this task.

G. SCHUSTER
Director-General
for Research, Science
and Education

Table of contents

Annexes

Introduction

The Commission of the European Communities is actively involved in energy research by programmes carried out in its establishments of the Joint Research Centre and by several other programmes (thermonuclear fusion, coal research, new energies etc.) carried out under contract by industry, universities and research institutes of its member countries.

This report is presenting the R & D work executed under contract in the framework of the "Energy Research and Development Programme" adopted by the Council of Ministers on 22nd August 1975, a programme dealing with the so-called "new energies".

After some general information on the programme's objectives, its management and administration, a detailed description of each of the contracts concluded before 30th May 1977 is given. This description, which constitutes the main part of this report (or "catalogue"), is subdivided into five chapters, following the scheme given below. Each chapter is preceded by a few pages presenting the main lines of research in the particular area considered.

I. The Council's programme decision

This decision, taken on 22nd August 1975 (*), adopts an energy research and development programme to be implemented under the responsibility of the Commission of the European Communities. It covers the following fields or sub-programmes (financial appropriations for four years are indicated between brackets):

1. Energy Conservation (11.38 m.u.a.)

   Subdivided into the following sectors:

   a - improved insulation of buildings;

   b - use of heat pumps;

   c - urban transport;

   d - residual heat recovery;

   e - materials recycling;

   f - production of energy from waste;

   g - evaluation of the specific energy consumption of equipment, processes and techniques;

   h - development of methods for storage of secondary energy.

(*) Official Journal No L 231 of 2 September 1975 (see annex I)

2. **Production and utilization of hydrogen** (13.24 m.u.a.)

    Subdivided into the following projects:

    A. Thermochemical production of hydrogen

    B. Electrolytic production of hydrogen

    C. Utilization of hydrogen

3. **Solar Energy** (17.5 m.u.a.)

    Subdivided into the following projects:

    A. Solar heat collectors and their application to dwellings

    B. Self-contained generating sets for the production of
        mechanical and/or electrical power

    C. Photovoltaic conversion

    D. Photochemical, photoelectrochemical and photobiological
        processes

    E. Photosynthetic production of organic matter

    F. Data network relating to solar radiation

4. **Geothermal Energy** (13 m.u.a.)

    Subdivided into the following projects:

    A. Acquisition and collation of existing and new geothermal
        data

    B. Improvement of methods of exploration

    C. Sources of hot water (low enthalpy)

    D. Steam sources (high enthalpy) and hot rocks

    E. Training of specialists

5. **Systems Analysis: Development of Models** (3.88 m.u.a.)

    Subdivided into the following actions:

    A. Static models (short term)

    B. Dynamic sector models (medium/long term)

The programme, which shall last from 1st July 1975 to 30th June 1979, with a financial outfit of 59 m.u.a., has been subdivided into two time phases of 1½ and 2½ years. At the end of the first phase, the programme content has been slightly reviewed*, without changing the financial appropriations. All contracts described in this report are emanating from the first phase.

The appropriations in the table below cover all the expenditure involved in implementing the programme, namely:

- expenditure relating to the Commission's financial obligations arising out of research contracts already concluded or still to be concluded;

- administrative operating expenditure;

- expenditure in respect of staff assigned to this work.

---

*Council Decision of 21 December 1976, Official Journal No L 10 of 13 January 1977.

TABLE OF FINANCIAL APPROPRIATIONS

(u.a.)

Objective (sub-programme)	FIRST PHASE			SECOND PHASE				Total Programme Allocation
	1.7.1975 (1)	1976 (2)	sub-total	1977	1978	30.6.1979	sub-total	
Energy Conservation	9,586	2,373,267	2,382,853	4,522,776	4,000,933	473,438	8,997,147	11,380,000
Production and utilization of hydrogen	2,000	2,327,203	2,329,203	5,415,695	4,814,099	681,003	10,910,797	13,240,000
Solar Energy	16,662	3,708,891	3,725,553	7,022,776	6,020,106	731,565	13,774,447	17,500,000
Geothermal Energy	2,742	2,747,999	2,750,741	5,213,678	4,512,096	523,485	10,249,259	13,000,000
Systems Analysis: Development of Models	11,558	745,533	757,091	1,626,176	1,021,149	475,584	3,122,909	3,880,000

(1) Commitments contracted

(2) Including appropriations for commitment outstanding from 1975

II. Implementation and supervision structures

Although the Commission largely calls upon assistance and advice from
Committees and individuals from the Member States, it bears the final
responsibility for the implementation of the programme. For its co-
ordination and management, a Division XII-0-2 "Energy R & D" has been
set up within the Commission's Directorate General XII "Research,
Science and Education".

Division XII-0-2 is headed by Dr. Ing. A. STRUB.

Each of the sub-programmes ("Objectives") is managed by a team, under
the conduction of the following programme co-ordinators:

Energy Conservation: Dr. H. EHRINGER

Production and utilization of hydrogen: Dr. H. MARCHANDISE

Solar Energy: Dr. A. STRUB (Deputy: Dr. W. PALZ)

Geothermal Energy: Dr. H. MARCHANDISE

Systems Analysis: Development of Models: Dr. E. RÖMBERG

For the technical and scientific guidance of each individual project
the Commission's services are assisted by recognized experts in the
different fields and disciplines required, the so-called "project
leaders" or "expert rapporteurs". They are either Commission officials
of the Joint Research Centre (JRC) or, more frequently, experts from
outside (universities, research institutes etc.) linked to the
Commission by an expert contract.

The actual list of project leaders and expert rapporteurs is given in
Annex II.

In order to assist the Commission in the proper implementation of the
programme and to ensure a good liaison with corresponding R & D work
carried out in the Member States, Advisory Committees of Programme
Management (ACPM) have been set up for each of the five sub-programmes.

Each ACPM contributes, in its advisory role, to the optimum execution
of the programme for which it has been set up. In this sense, it
takes part in such tasks as:

- definition of detailed steps related to programme implementation;

- selection of research proposals;

- evaluation of results;

- selection of "project leaders" or "expert rapporteurs".

Each Committee is composed of national delegations (up to 3 represent-
atives per Member State) and a Commission delegation.

The names of the members of the five ACPM are listed in Annex III.

For the overall steering of the programme the Commission is assisted
by the Scientific and Technical Research Committee(CREST)*.   CREST
advises the Commission and the Council and has to deal with all
questions of science and technology falling within the EC treaty.   It
ensures that there is a general consistency and mutual co-ordination
between EC and national programmes.

In order to do so, CREST is backed by a number of sub-committees among
which the Energy R & D sub-committee ("CREST-E") is specially concerned
with the present programme.   This sub-committee has, besides many other
tasks, to assist the Commission in the elaboration, supervision and
co-ordination of all its energy R & D programmes.   The members of CREST
and of its sub-committee on energy are listed in Annexes IV and V
respectively.

III.   Status of implementation

About 600 research proposals were received in the first phase of the
programme up to now (30/5/77).   Their assessment and evaluation has
led to the conclusion of 191 contracts.

The overall contractual situation is shown in the following table:

_____

*Comité de la Recherche Scientifique et Technique.

Objective (sub-programme)	Number of contracts	Total cost u.a.	EC contribution u.a.	Number of proposals under negotiation
Energy conservation	27	3,572,940	1,889,334	22
Production and utilization of hydrogen	29	3,393,638	1,828,235	-
Solar Energy	76	7,097,115	2,832,051	5
Geothermal Energy	49	4,016,940	2,043,514	3
Systems Analysis: Development of Models	10	652,424	602,424	-
TOTAL	191	18,733,057	9,195,558	30

It can be seen from this table that the average EC contribution is
about 49% and that the average cost of a research contract amounts
to 98,100 u.a.

A more detailed financial breakdown can be found in the individual
chapters devoted to each sub-programme.

IV. Information for future proponents

1. Invitation to tender - Submission of Proposals

The Commission normally seeks to attract research proposals for
contracts by issuing calls for tenders for specific subjects of
its interest.   Where the nature of the subject may interest a
great many proponents, the call for tender is published in the
Official Journal of the Communities.

Where the nature of the research is only of interest or within
the range of a small number of specialized research institutes
the invitation to tender is limited in agreement with the ACPM.

Research proposals should only be introduced to the Commission
with reference to an invitation for tenders and exclusively by
using the special EC form for submission of proposals.   This
form can be requested at the following address (indicating the
sub-programme and the call for tenders to which it refers):

> Commission of the European Communities
> Energy R & D Programme
> General Directorate XII
> 200 rue de la Loi - B-1049 Brussels

2. Research contracts

Up to now, all contracts concluded are of the "individual" type,
i.e., between the Commission and one partner.   Multiple contracts
between the Commission and several partners could also be made up
if the subjects lend themselves to it.

In principle, the cost of the research to be carried out in each
contract is shared between the Commission and the contractor.
This latter must be able to prove that the share not funded by
the Commission (50% in standard cases) is available either from
the contractor's own budget or from a third party.

3. Dissemination of information, property rights

The provisions for dissemination of information and patents
resulting from the research carried out are laid down in a special
annex to each contract.   The standard text of these provisions as
normally attached to each research contract is given in Annex VI.

Energy conservation

Energy losses, occurring both in the transformation of primary energy
into usable energy and in final consumption, are estimated in the
Community at about 54% of total requirements.   The reduction of these
losses has become a major objective, with particular emphasis on the
following sectors:

- domestic and tertiary (losses estimated at 55%)
- transport   (losses estimated at 83%)
- industry   (losses estimated at 45%)
- industry and energy
  production   (losses estimated at 60%)

The reduction of these losses can be achieved mainly by:

(a) a more rational use of energy, based on the reduction of non-
     essential consumption and on the improvement of yield, to be
     carried out primarily by means of measures ranging from public
     information to the study of administrative previsions and
     technical standards;

(b) research and development projects aimed at:

     - improving processes, equipment, etc., on the basis of
       existing technology;

     - studying and developing new technologies.

The EC energy conservation R & D programme concentrates on these
areas in which R & D activity is particularly necessary to improve
the technologies in use at present.   In addition, its aim is also
to stimulate technological innovation in new fields.

The programme is sub-divided in eight sectors.   For each of these
sectors the general scope of work is given hereunder as well as the
subjects covered by research contracts selected during the first
programme phase.[1]

---

[1]In the second phase, R & D on heat pumps (B), recovery of waste
heat (D) and storage of secondary heat (H) will be emphasised.

Sector a: Improved insulation of buildings

  Scope: Ways will be studied to improve the insulation properties
          of both transparent and non-transparent materials used in
          buildings.

  Fields_covered: Mainly the study of transparent materials (windows).

Sector b: Use of heat pumps

  Scope: Development of heat pumps for heating and cooling purposes
          and for recovery of waste heat in industry.

  Fields_covered: - Development of heat pump technology, modular
          systems and high temperature heat pumps producing heat at
          around 120°C at the condenser;

                   - Exploration of the potential of soil for heat
          extraction by heat pumps and an evaluation of the
          environmental impact;

                   - Comparison of different heat pump systems.

Sector c: Urban transport

  Scope: Improvement of the energy efficiency of vehicles.

  Fields_covered: - Study of part load behaviour of spark ignition
          engines;

                   - Development of new types of diesel engines
          (variable chamber geometry, no coolant engine);

                   - Comparison of energy consumption of different
          engines (e.g., diesel, gasoline) under different conditions
          of use.

Sector d: Residual heat recovery

  Scope: R & D on components and systems for the recovery and
          transport of waste heat.

  Fields_covered: - Studies on the recovery of waste heat in the
          glass and coke industry;

                   - The use of low temperature Rankine engines to
          convert waste heat into work (e.g., in trucks and industrial
          plants;
                   - Comparative study of two types of heat recovery
          systems (heat pipes, rotating generator).

Sector e: Materials recycling

Scope: The development of processes which recycle materials with a much lower energy requirement than processes producing new materials (including economic assessments).

Fields covered: Recycling of plastics extracted from urban and industrial waste.

Sector f: Production of energy from waste

Scope: Study of the waste available and of the best ways to process it for energy production (e.g. sorting, combustion).

Fields covered: Fluidized bed combustion and gasification of low grade materials such as coal refuse, straw, wood, urban wastes.

Sector g: Evaluation of the specific energy consumption of equipment, processes and techniques

Scope: Analysis of the energy content of products and of the energy consumption of processes with the aim of optimizing the production processes from the standpoint of energy use.

Fields covered: - Development of a methodology for optimizing energy use in industrial plants;

- Study of energy consumption in electric water heating (demand pattern, technology);

- Reduction of energy consumption in farm cooling of milk;
- Study of the variable steam extraction rates in a steam turbine.

Sector h: Development of methods for storage of secondary energy

Scope: Study and development of systems and components for:

- Heat storage in the form of sensible and latent heat and chemical storage;

- Storage of electricity: fuel cells, batteries, flywheels, compressed air, storage lakes;

- etc.

Fields covered: - Development of an electrolyte for a sodium-sulphur high temperature battery;

- Use of molten salts for heat storage.

SUMMARY AND BREAKDOWN
OF FUNDING

Objective: Energy conservation

Sector	Number of contracts(*)	Total cost u.a.	E.C. contribution u.a.	Number of proposals under negotiation
a	3	339,440	169,720	2
b	6	441,223	217,067	2
c	3	497,493	244,620	1
d	5	667,045	343,312	5
e	1	319,670	159,835	1
f	3	396,319	198,160	0
g	5	596,420	367,417	9
h	1	315,330	189,203	2
Total	27	3,572,940	1,889,334	22

(*)Signed both by the Commission and the contractor or sent for
signature to the contractor.

Sector a:

Improved insulation of buildings

COMMISSION OF THE EUROPEAN COMMUNITIES ENERGY R & D PROGRAMME Objective: Energy conservation	Project or Sector: Improvement of insulation in buildings

Title: Energy saving in housing by infrared reflective and spectral selective finishing.	

Duration: 10 months Period: 15/10/1976 - 15/8/1977	Contract No: 144-76 EENL Project No:

Contractor: Institute of Applied Physics TNO-TH

Address: P.O. Box 155, Delft, Netherlands

Head of Project: Dr. ir. J. de Jong

Description of research work

I. Objectives (aims)

This project concerns energy saving in housing by infrared reflective and spectral selective finishing. It's aim is to prove that heat reflective finishing can lead to substantial energy savings with respect to the heating of buildings and dwellings. Energy conservation with the reflective insulation method is making use of a lowering of the desired air temperature as a result of an increase in the mean radiant temperature. The following economic advantages can be expected:

- saving of energy against relative low cost, which is also applicable in existing dwellings;

- no loss of volume of the dwellings.

2. Work Programme

The work programme consists of a two-part study programme:

(a) Classification of products and materials with heat reflective properties. For various materials such as aluminium, steel, glass, paint etc., the following properties will be considered:

- reflection and absorption as a function of the wave length for visible light and infrared radiation;

- mechanical and chemical properties;

- light stability.   ../..

Total cost:
  Fl. 140,780
  u.a. 38,890
	E.C. Contribution: 50%
	  Fl. 70,390
	  u.a. 19,445

COMMISSION OF THE EUROPEAN COMMUNITIES ENERGY R & D PROGRAMME **Objective:** Energy conservation	**Project or Sector:** Improvement of insulation in buildings

2.   Work Programme   cont..:

(b)   Determination of energy saving:

These determinations will be performed for two types of living room, each with two heating systems (radiation heating versus air heating) and various types of heat reflective finishing (e.g. diffuse reflecting versus directional reflecting).

The following points will be determined:

- the rise in the mean radiant temperature and its effect on the air temperature, from the point of view of thermal comfort for human beings;

- the enlargement of the heat resistance of the various parts of the enveloping construction;

- cost benefit analyses.

3.   Status

Work in progress.

COMMISSION OF THE EUROPEAN COMMUNITIES ENERGY R & D PROGRAMME Objective: Energy Conservation	Project or Sector: Improvement of insulation in buildings

Title:

A technical economic study of selective glazing obtained by
coating with a transparent conductive deposit.

Duration: 12 months Period: 10/5/77 - 10/5/77	Contract No: 201-77 EEF Project No:

Contractor: Centre d'Etudes Nucléaires de Granoble (C.E.N.G.)

Address: B.P. 85, Centre de Tri, F-38041 Grenoble Cedex, France

Head of Project: M. G. Blandenet

Description of research work

I. Objectives (aims)

A technical economic study of selective glazing obtained by coating with
a transparent conductive deposit.   Glazing of this type is intended to
improve the heat insulation of buildings by reducing losses due to infrared radiation.   Such losses account for 30-40% of the total losses from
a building.

The study will be based on the PYROSOL process developed by CENG for
coating glass panels by pyrolisis of aerosols formed by ultrasonic
atomization.

The aim of the research is to carry out a systematic study which will
enable a comparison to be made of the cost and quality of each type of
product, and the best quality/price ratio to be determined.

2. Work Programme

Comprises two sections:

(a) Technical

- optimization of indium oxide;
- study on the addition of SnC_{12}, SnC_{14}, diacetyl butyl tin,
  fluorine and antimony to tin oxide;
- characterization: physical:- examination of electrical and
                                 optical properties;
                              - aging by stoving and cyclic heating
                                by the Joule effect;   ../..

Total cost:
   FF.   379,536
   u.a.   68,336
	E.C. Contribution: 50%
	   FF. 189,768
	   u.a. 34,168

COMMISSION OF THE
EUROPEAN COMMUNITIES

ENERGY R & D PROGRAMME

Objective: **Energy Conservation**

Project or Sector:

**Improvement of insulation
in buildings**

2. Work Programme cont/..

- chemical: resistance to chemical agents by prolonged immersion
in soda, nitric acid and detergent solutions.

(b) **Economic aspect**

A cost estimate will be made on the basis of carefully executed
tests where account is taken of the small dimensions of the backings.

In view of the technical facilities employed, the time factor will
be disregarded.

3. **Status**

Work in Progress.

COMMISSION OF THE EUROPEAN COMMUNITIES ENERGY R & D PROGRAMME Objective: Energy Conservation	Project or Sector: Improvement of insulation in buildings

| Title:

   Selective optical coating on plastic sheet for inexpensive radiation insulation of visible window. ||

Duration: 26.5 months Period: 15/4/77 - 30/6/79	Contract No: 212-77 EEUK Project No:

Contractor: Loughborough University of Technology, Department of Physics

Address: Leics, 3 TU, United Kingdom

Head of Project: Dr. R.P. Howson

Description of research work

I.   Objectives (aims)

The aim of the research is to investigate novel vacuum techniques for creating films of indium oxide and similar elemental and mixed oxides which are suitable for inexpensive production technology which principally depends upon the immediate substrate being a plastic sheet.

Films of indium tin oxide have proved to have the desired optical properties when produced by chemical techniques onto heated glass surfaces. Conventional vacuum techniques at lower temperatures have not been so successful.

Novel vacuum techniques using reaction on the substrate surface have proved successful with simiconductor and oxide films and the programme is intended to extend the range of these methods using evaporation within an active gas created by a radio frequency discharge.

Reactive evaporation of the metal element in a low pressure triode discharge promises to be a rapid technique which could be done at a vacuum pressure suitable for production equipment and which could use low temperature substrate materials such as plastic sheet.

Two techniques will be investigated whixh could be applicable to a commercial high speed batch evaporation system which will have a relatively high residual gas pressure: the process of reactive evaporation and vacuum anodisation.

Total cost:
  £   96,755
  u.a. 232,214
	E.C. Contribution: 50%
	  £ 48,378
	  u.a. 116,107

COMMISSION OF THE EUROPEAN COMMUNITIES ENERGY R & D PROGRAMME Objective: **Energy Conservation**	Project or Sector: **Improvement of insulation in buildings**

2. Work Programme cont/..

(a) Development of the film production technique;

(b) Vacuum anodisation technique

- direct current plasma in a vacuum system

- radio frequency plasma in a vacuum system;

(c) Reactive evaporation technique

- direct current triode plasma

- radio frequency plasma in a vacuum system

(d) Report.

3. Status

Work in progress.

Sector b:

Use of heat pumps

COMMISSION OF THE EUROPEAN COMMUNITIES ENERGY R & D PROGRAMME Objective: Energy conservation	Project or Sector: Use of heat pumps

Title:

Analysis of the factors which determine the COP of a heat pump, and a feasibility study on ways and means of increasing it.

Duration: 16 months Period: 7/3/1977 - 7/7/1978	Contract No: 142-76 EEDK Project No:

Contractor: European heat pump consultors

Address: Rosenk 22B - 2860 SøBorg, Denmark

Head of Project: Mr. Fordsmand Technical responsibility
Mr. A. Eggers-Lura Administrative responsibility

Description of research work

I. Objectives (aims)

The aim of this contract is the development and testing of a prototype of a second generation industrial heat pump of a modular Fordsmand type, using the earth as a heat source. The information and results, plus all the facts concerning this heat source, will be presented in such a way that a realistic comparison with analyses to be made by TNO and the activities of VEW, is possible.

2. Work Programme

The work programme is divided into two sub-projects.

(a) Presenting the experimental and empirical experiences on heat pumps over the past twenty years, and more especially, over the past two years. The presentation of the measurement results obtained during the past two years of heat pumps, using the earth as the heat source.

(b) The presentation and description of experiences with the first generation heat pump of a modular type in accordance with the concepts of Mr. Fordsmand. The development and testing of a prototype second generation heat pump, based on the concept of Mr. Fordsmand. The heat pump is electrically driven and can be used for a wide range of heat sources and storage with high efficiency. The heat pump is specially adapted for the heating of a one-family house, and for the production of domestic hot water.

3. Status Work in progress

Total cost: Dk. 531,500 u.a. 70,867	E.C. Contribution: 45% Dk. 239,175 u.a. 31,890

COMMISSION OF THE EUROPEAN COMMUNITIES ENERGY R & D PROGRAMME Objective:　Energy conservation	Project or Sector: Use of heat pumps

| Title:

　　　Industrial application of high temperature heat pumps
　　　driven by prime movers. ||

Duration:　　12 months Period:　　　1/3/1977 - 1/3/1978	Contract No:　175-77　EEUK Project No:

| Contractor:　International Research and Development Co. Ltd. (I.R.D.)

Address:　　Fossway, Newcastle-upon-Tyne,　NE6 2YD,　England

Head of Project: ||

Description of research work

I. Objectives (aims)

The aim is to design a heat pump system for operation at a condensing temperature of 120°C, using as a prime mover a gas engine.　Recovery of waste heat will take place from the exhaust and water jacket and eventually from the oil cooler.　The use of heat pumps is required in industrial processes, where there is a need for considerably higher condensing temperatures, in the range of 100°C and higher, and when large quantities of heat are required.

2. Work Programme

The programme is divided into two stages:

Stage 1:　Specification and design of the system with an economic
　　　　　analysis.　A detailed set-up of the programme of Stage 2 will
　　　　　be made.

Stage 2:　Procurement, installation and tests.

Following Stage 1, an evaluation of the results and a detailed evaluation of the programme Stage 2 will be executed and a decision will be taken by the Commission regarding the financial support for Stage 2.

3. Status

Work in progress

Total cost: 　　£　9,361 　u.a. 22,468	E.C. Contribution:　　50% 　　£　　4,681 　u.a.　11,234

COMMISSION OF THE EUROPEAN COMMUNITIES ENERGY R & D PROGRAMME **Objective:** Energy conservation	**Project or Sector:** Use of heat pumps

Title:

Study of the specific energy consumption of various heat pump systems and examination of the effects on the environment of the operation of heat pumps.

Duration: 16 months **Period:** 1/12/76 - 31/3/78	**Contract No:** 191-77 EED **Project No:**

Contractor: V.E.W.
Vereinigte Elektrizitätswerke Westfalen A.G.

Address: Ostwall 51, D-4600 Dortmund

Head of Project: Herrn Dipl. Ing. P. Müller

Description of research work

I. Objectives (aims)

Using a heat pump, heat can be extracted from water, from the ground or from air by the use of mechanical energy.

Inevitably, the extraction of heat from ground and surface water, from the ground and from air brings about physical changes in these heat sources.   The object of this study is to quantify these changes.   In this way it should be possible, circumstances permitting, to determine the adverse effects on the environment.

2. Work Programme

The object of the work under contract is to make a metrological study of the effects of the extraction of heat by heat pump systems on water, ground and air.

Ground water

Observations are to be made on existing installations using, in all, about thirty observation wells arranged radially around the extraction and reinjection bores.   The purpose is to observe the extent of the ground-water drawdown cone which occurs at the withdrawal point and that of the reinjection bore.   At the same time, the relative temperatures of various soil types and the rate of flow of the groundwater are to be

../..

Total cost: DM. 697,442 u.a. 190,558	**E.C. Contribution:** 50% DM. 348,721 u.a. 95,279

COMMISSION OF THE EUROPEAN COMMUNITIES	Project or Sector:
ENERGY R & D PROGRAMME	Use of heat pumps
Objective: Energy conservation	

2. Work Programme cont...

Ground water cont...

recorded.   Measurements are also to be taken of the energy consumed by the extraction and reinjection installations and of the electric current used by the heat pump plant as a whole.

Earth

Heat pump installations, using the ground as the heat source, are to be used to investigate the relative temperatures of various soil types at the level of the heat exchange surfaces, on a plane at right angles to the heat exchange surfaces and in the soil not reached by the water. In addition, the depth of the heat exchange surfaces and the horizontal distance between heat exchange ducts will be varied.   Metrological studies are to be made of the electrical energy consumption of the heat pump installations.

Air

Heat pump installations, using air as the heat source, are to be used, depending on the amount of air discharged, to investigate the sound pressure level and the extent of the cold air plume depending on climatic conditions.   The yearly performance figures of the heat pump installation, with and without additional energy, and the practical operating conditions of various de-icing systems are also of interest.   The current consumed by the ventilator and the system as a whole is also to be measured.

3. Status

Work in progress

COMMISSION OF THE EUROPEAN COMMUNITIES ENERGY R & D PROGRAMME Objective: Energy Conservation	Project or Sector: Use of heat pumps

Title:

Energy saving heat pump systems coupled with sewage renovation systems.

Duration: 12 months Period: 1/1/77 - 31/12/77	Contract No: 211-77 EED Project No:

Contractor: M.A.N. (Maschinenfabrik Augsburg Nürnberg AG)

Address: Dachauerstrasse 667, D-8000 München 50, Germany

Head of Project: Ing. M. Simon

Description of research work

I. Objectives (aims)

Energy saving by compressor/absorber heat pumps driven by internal combustion engines combined with water purification and treatment plants for use in sewage systems and polluted waters.
Aim of the project: feasibility study and project definition. The contract covers the feasibility study and the definition of the main project.

2. Work Programme

The project centres on a comparative examination of various heat pump drive systems, in particular the combustion engine systems, since these are expected to have a primary energy efficiency of up to 300% as against approximately 70% with conventional heating and 30% with all-electric heating.

Attention will also be given to the integrated energy balance of certain heating systems and complementary subsystems for peak load service. Their integration into the sewage purification plants i.e., the generation of fuel with separated waste products as well as the extraction of heat forms part of the study.

Heat pump specification

The heat will be used for heating purposes. Power spectrum and range of applications: from the kW to the MW range, and the optimum driving machinery for the purpose.

3. Status   Signature

Total cost: DM. 299,963 u.a. 81,957	E.C. Contribution: 50% DM. 149,982 u.a. 40,978

COMMISSION OF THE EUROPEAN COMMUNITIES ENERGY R & D PROGRAMME Objective: **Energy Conservation**	Project or Sector: Use of heat pumps

Title:

Development of an absorption heat pump of compact design for domestic heating.

Duration: **12 months** Period: **15/2/77 - 15/2/78**	Contract No: **214-77 EED** Project No:

Contractor: **Battelle Institut**

Address: **Am Römerhof 35, Postfach 900160,**
D-6000 Frankfurt/M 90, Germany

Head of Project: **Mr. G. Oelert**

Description of research work

I. Objectives (aims)

At the present time, electrically operated compressor heat pumps are being used with technical success for domestic heating purposes. These are plants that draw the heat they require from the soil, from ground-water or from the ambient air.

The operating costs depend on many factors. Owing to the high cost of electricity the running costs are much the same as those of oil or gas-fired central heating, even where the compression heat pump has an approximate coefficient of performance of E_{eff}= 4. The installation costs for heat pumps, however, are higher than those of comparable conventional central heating systems.

Interesting possibilities of real energy-saving are offered by a heat pump which derives its operating energy from direct combustion of non-polluting fuels and which contains as far moving parts as possible.

An essential feature of this heat pump (apart from the fact that it is driven by primary energy) is that its performance is no less efficient than that of a conventional boiler system, even driven the brief spells wh when the temperature of the heat source i.e., ambient air, is extremely low. Hence, unlike the electrically driven heat pump, it does not need to be combined with a conventional boiler to form a "bivalent system".

../..

Total cost: DM. **120,545** u.a. **32,936**	E.C. Contribution: **50%** DM. **60,273** u.a. **16,468**

COMMISSION OF THE EUROPEAN COMMUNITIES ENERGY R & D PROGRAMME Objective: **Energy Conservation**	Project or Sector: Use of heat pumps

2. Work Programme

A study is in progress on the possibility of using natural gas or oil directly for the operation of an absorption heat pump, and an economic evaluation is being made by comparison with the present state of the art in oil and gas heating.

(a) Design

- Development of the drive concept;

- Technical assessment of tried absorption systems and main system components;

- Market research on existing supply outlets.

3. Status

Work in progress.

COMMISSION OF THE EUROPEAN COMMUNITIES ENERGY R & D PROGRAMME Objective: Energy Conservation	Project or Sector: **Use of heat pumps**

Title:

      Investigation about using the soil as a natural heat source for heat pumps.

Duration: **20 months** Period:	Contract No: **231-77** EEC Project No:

Contractor: **Centraal Technisch Instituut T.N.O.**

Address: **Laan van Westenenk 501, Apeldoorn, Netherlands**

Head of Project: **Ir. H. van der Ree**

Description of research work

I. Objectives (aims)

Heat pump applications in general and the use of the soil as a natural heat source for this apparatus in particular represent a new technique, related to conventional heating systems. The soil may also be used as a heat sink in summer and in such a way a seasonal storage system is created. Economic advantages are expected as this application represents an energy saving technique.

The crucial problem in applying heat pumps for the heating of houses is finding a suitable natural heat source for the extraction of low-grade heat. Common heat sources such as ground water and outside air have specific drawbacks, which are not encountered when using the soil.

The aim of the project is evaluating the soil as a natural heat source for heat pumps for the heating of houses. Among the points to be investigated are how much heat can be stored in and extracted from the soil during an average summer and winter.

2. Work Programme

(a) Studying relevant literature; collecting basic data;
(b) Developing mathematical model;
(c) Optimising buried heat exchanger;
(d) Heat transfer calculations;
(e) Reporting and calculating thermal behaviour of heat pump system;
(f) Technical and economical evaluation
     Final reporting.

3. Status   Signature

Total cost:
   Fl.   153,622
   u.a.   42,437
	E.C. Contribution: 50%
	   Fl. 76,811
	   u.a. 21,218

Sector o:

Urban transport

COMMISSION OF THE EUROPEAN COMMUNITIES ENERGY R & D PROGRAMME Objective: Energy conservation	Project or Sector: Urban transport

Title:

Analysis of the part-load operating behaviour of spark-ignition engines.

Duration: 18 months Period: 31/12/1976 - 31/6/1978	Contract No: 122-76 EEF Project No:

Contractor: Institut Français du Pétrole

Address: 1-4 Avenue du Bois-Préau, B.P. 18
                      92502 Rueil Malmaison, France

Head of Project: Mr. P. Eyzat

Description of research work

I. Objectives (aims)

This contract sets out to analyse the part-load operating behaviour of spark-ignition engines and to define optimum approaches as regards energy efficiency.   The fact that fleets of vehicles, and the conditions under which they are used and driven in Europe, are homogeneous warrants the expectation that the results obtained will find widespread application.

2. Work Programme

This comprises a test programme on a single-cylinder engine which involves:

- tests under variable speed and load conditions;

- thermodynamic analysis of part-load operating behaviour;

- parametric study of the influence of the law of combustion on efficiency;

- influence of gas/wall thermal exchanges on efficiency;

- tests under conditions of variable intake mixture composition and thermodynamic analysis;

- tests under conditions of variable compression ratio and experimental study of the influence of other parameters, such as, ignition characteristics and lubricant temperatures.

                                                                ../..

Total cost:
  FF.   374,460
  u.a.   67,422
	E.C. Contribution: 50%
	  FF. 187,230
	  u.a. 33,711

COMMISSION OF THE EUROPEAN COMMUNITIES ENERGY R & D PROGRAMME **Objective:** Energy conservation	**Project or Sector:** Urban transport

2. Work Programme cont...

On the basis of the results obtained a plan of study for a multi-cylinder engine will be adapted and carried out.

3. Status

Work in progress.

COMMISSION OF THE EUROPEAN COMMUNITIES ENERGY R & D PROGRAMME Objective: Energy conservation	Project or Sector: Urban transport

Title:

Preliminary study of the adaptability of a variable geometry chamber to a diesel motor vehicle.

Duration: 18 months Period: 1/7/1976 - 31/12/1977	Contract No: 123-76 EEE Project No:

Contractor: Institut Français du Pétrole

Address: 1-4 Avenue du Bois Préau, B.P. 18
92502 Rueil Malmaison, France

Head of Project: Mr. P. Eyzat

Description of research work

I. Objectives (aims)

Preliminary study to assess whether the variable-geometry combustion chamber concept can be adapted to a motor vehicle diesel engine, with the aim of achieving an energy saving from the reduction in specific consumption of motor fuel. This aim will be achieved:

- directly; through a reduction of heat transferred to the walls, attributable to the principle on which the combustion chamber works;

- indirectly; by enabling the engine to work with a lower volumetric ratio than any ratio adopted in traditional approaches.

2. Work Programme

The research work divides into two distinct areas: the first relating to single-cylinder and the second to multi-cylinder engines. This contract is limited to the single-cylinder area and comprises:

- a preliminary design, namely, the definition of the principal characteristics of the combustion chamber;

- study and production of a prototype cylinder head and piston;

- tests and adaptation of prototype injectors;

../..

Total cost:
 FF.   1,929,845
 u.a.   347,469
	E.C. Contribution: 50%
	  FF. 964,923
	  u.a. 173,735

COMMISSION OF THE EUROPEAN COMMUNITIES ENERGY R & D PROGRAMME **Objective:** Energy conservation	**Project or Sector:** Urban transport

2. Work Programme cont...

- measurement of the characteristics and influence of the main parameters;

- study of behaviour at start-up and also under conditions of over-richness.

The second area will only be undertaken with the aid of one or more manufacturers and on the basis of the results obtained from the first area.

3. Status

Work in progress.

COMMISSION OF THE EUROPEAN COMMUNITIES ENERGY R & D PROGRAMME Objective: Energy conservation	Project or Sector: Urban transport

Title:

No coolant diesel engine.

Duration: 33 months Period: 1/10/76 - 30/6/79	Contract No: 143-76 EEIR Project No:

Contractor: University College Dublin

Address: Merrion Street, Dublin 2, Ireland

Head of Project: Prof. Seamus Timoney

Description of research work

I. Objectives (aims)

The project covers the construction of a double-piston single cylinder 300-500 cc. engine with ceramic piston and liner, and also the testing of the engine in order to acquire data for the design of a second generation experimental engine. The engine can be considered as the forerunner of a multi-fuel engine of medium power and speed, giving about 100-200 kw/cylinder.

2. Work Programme

The work programme is divided into two stages. The first stage consists of the design, manufacture of subassemblies and the evaluation of the possibilities of success.

The second stage consists of the mounting, testing under various conditions and the stripping down of the engine for examination. After the rebuilding of the engine, the research consists of increasing the power output.

3. Status

Work in progress.

Total cost: £ 34,420 u.a. 82,602	E.C. Contribution: 45% £ 15,849 u.a. 37,174

Sector d:

Residual heat recovery

COMMISSION OF THE EUROPEAN COMMUNITIES ENERGY R & D PROGRAMME Objective: Energy conservation	Project or Sector: Recovery of residual heat

Title:

    Convection furnace for preheating glass compounds in the form of granules.

Duration: 12 months Period: 31/12/1976 - 31/12/1977	Contract No: 169-76 EET Project No:

Contractor: Sté. Bertin et Cie.

Address: B.P. 3 - 78370 Plaisir, France

Head of Project: Mr. G. Dahan

Description of research work

I. Objectives (aims)

Research and development of a convection furnace using waste gases as the means of preheating a glass compound put up in the form of granules. The granules are intended as material for feeding furnaces for float glass or the manufacture of expanded glass beads. The latter constitute a novel insulating product which can be added to concrete or used as a filling material.

2. Work Programme

- Definition of the technical approaches which enable the glass compound granules to be preheated by the percolation of waste gases;

- Study of problems of handling at the inlet and outlet of the furnace;

- Determination of the aerodynamic losses incurred with various types of granules in terms of various parameters:

 - size of the test furnace;

 - design and construction of the test furnace;

 - experimentation and utilization of the results with the aim of defining all the parameters which enable the size of an industrial furnace to be optimized.

        ../..

Total cost:
  FF.   640,091
  u.a. 115,249
	E.C. Contribution: 50%
	  FF. 374,414
	  u.a. 67,413

COMMISSION OF THE EUROPEAN COMMUNITIES ENERGY R & D PROGRAMME **Objective:** Energy conservation	**Project or Sector:** Recovery of residual heat

2. Work Programme cont...

On the strength of the results obtained a second stage could be contemplated involving the design of a pilot furnace with a capacity of 0.2 - 2 t/h which would enable an industrial furnace with a capacity of 30 t/h to be designed and built.

3. Status

Work in progress.

COMMISSION OF THE EUROPEAN COMMUNITIES ENERGY R & D PROGRAMME Objective: Energy conservation	Project or Sector: Recovery of residual heat

Title:

Study of the possibilities of recovering lost heat in thermal installations.

Duration: 24 months Period: 15/12/1976 - 15/12/1978	Contract No: 173-76 EEF Project No:

Contractor: CERCHAR

Address: 33 rue de la Baume, 75008 Paris

Head of Project: Mr. Busso (Chef du groupe "Recherches Thermiques")

Description of research work

I. Objectives (aims)

The proposed aim is to search for technical and economical possibilities of recovering lost heat in traditional coking-plants, where these plants are taken as examples of establishments having high thermal energy consumption. The idea is to find out, among losses capable of measurement, those which could be reduced, either in the short-term by major changes in plant operation, or in the long-term by more radical changes in technique, and perhaps even by completely changing certain procedures used in the manufacture of coke. The additional amount of coking-plant gas available for transfer to thermal installations or to the coal by-products industry, will correspond to the amount of heat recovered.

2. Work Programme

The programme, as suggested, consists of two stages:

(a) the setting-up of a survey of energy consumption and losses (heat, electricity), and the composing of a programme of more or less long-term priority action, with suggestions for pilot experiments on an industrial scale;

(b) experimental research to find a way of conserving the heat from gas distillation, as almost $\frac{1}{4}$ of the energy used to carbonize the coal mixtures is carried away by this gas.

3. Status Work in progress.

Total cost: FF. 705,945 u.a. 127,106	E.C. Contribution: 50% FF. 352,973 u.a. 63,553

COMMISSION OF THE EUROPEAN COMMUNITIES ENERGY R & D PROGRAMME Objective: Energy conservation	Project or Sector: Recovery of residual heat

| Title:

Comparative study of rotating regenerators and heat pipe heat exchangers. ||

Duration: 15 months	Contract No: 184-77 EEUK
Period: 25/5/1977 - 25/8/1978	Project No:

Contractor: International Research and Development Co. Ltd., I.R.D. Address: Fossway, Newcastle upon Tyne, NE6 2YD, England Head of Project: Mr. B. Forster

Description of research work

I. Objectives (aims)

The aim of the research is to assess the performance (technically and economically) of two forms of heat recovery systems, heat pipes and rotating regenerators. The heat pipes and rotating regenerators will be applicable to gaseous exhausts, particularly those containing contamination (moisture, dust or oil).

The work will lead to the provision of relevant financial and technical data on the optimum methods of recovering waste heat from processes where hot air/gas is used as the heating medium with particular emphasis on the quantification of contaminants on the performance.

2. Work Programme

The programme covers the testing of a rotating regenerator and a heat pipe heat exchanger at an ICI plant. It is divided into four stages:

(a) manufacture and testing of a heat pipe heat exchanger at IRD;

(b) procurement and testing of a rotating regenerator in a Brattice-type oven with a heat recovery capability of 20 to 40 kw;

(c) testing of a heat pipe heat exchanger in a Brattice-type oven;

(d) comparison of performance data.

3. Status   Work in progress.

Total cost:
    £   32,541
    u.a.   78,098
	E.C. Contribution: 50%
	    £ 16,271
	    u.a. 39,050

COMMISSION OF THE EUROPEAN COMMUNITIES ENERGY R & D PROGRAMME Objective : Energy conservation	Project or Sector: Recovery of residual heat

Title:

Experimental Rankine cycle engine designed for utilisation of low temperature, low pressure heat.

Duration: 17 months Period: 1/7/1976 - 1/12/1977	Contract N°: 196-76-7 EEI Project N° :

Contractor: FIAT

Address: Strada Torino, 50   I - 10043 Orbasssano

Head of Project: Mr. Prof. Businaro

Description of research work:

1.   Objectives (aims)

Experimental Rankine cycle engine (95 kW) to use low temperature; low pressure heat. Primary aim of this contract is to develop an engine to transform into mechanical or electrical energy the waste heat contained in heat engine exhaust gases and industrial exhaust, or the energy at non high temperature obtainable through solar collectors.

The recovery of the thermal energy contained in heat engine exhaust gases by means of this engine affords fuel economies of about 20 % in the case of regenerative gas turbines (gas around 350° C) and over 10 % for Diesel engines. Emission levels will be reduced by an amount equal to the gains in efficiency.

2.   Work Programme

The work programme consists of three main activities as follows :

a) a laboratory investigation on the compatibility and chemical stability of Freon under simulated operational conditions

b) the development of a pilot unit and of the relevant test equipment

c) experimental investigations on the unit under scheduled operation conditions.

3.   Status

Work in progress.

Total cost: Lit. 162,000,000 u.a. 259,200	E.E.C. Contribution: 50 % Lit. 81,000,000 u.a. 129,600

COMMISSION OF THE EUROPEAN COMMUNITIES ENERGY R & D PROGRAMME Objective : Energy conservation	Project or Sector: Recovery of residual heat.

Title:

Recovery of heat, in ovens for fining ceramics, for mechanical or electricity production.

Duration: 30 months	Contract N° : 198-77 EEI
Period: 1/1/1977 - 30/6/1979	Project N° :

Contractor: Gemmindustria S.N.C.
Dipartimento lavorazioni speciali

Address: Via Falcone, 7
I - 20100 MILANO

Head of Project: Prof. Ennio Macchi

Description of research work:

1. Objectives (aims)

Only a part of the heat from the tunnel oven producing ceramics is used to preheat and dry the clays before firing. The aim of the research is to recover the heat lost (200 to 250° C) by using a Rankine cycle to produce mechanical energy for the auxiliary devices in the facility itself (fans, crushers, compressed air-supply, ...). The following special problems will be examined : thermodynamic properties and characteristics, corrosion, particularly that caused by fluorine, wear, fouling of the heat-exchanger, etc.

2. Work programme

- data collection
- selection of transfert fluid
- design and construction of facility
- installation of heat exchanger in the oven, test without turbine, data collection
- installation of turbine, long term test
- dismantling and inspection of all components
- reassembly after correction have been carried out, followed the industrial-scale operation.

3. Status

Work in progress.

Total cost: Lit. 54,620,000 u.a. 87,392	E.E.C. Contribution: 50 % Lit. 27,310,000 u.a. 43,696

Sector e:

Materials recycling

COMMISSION OF THE EUROPEAN COMMUNITIES ENERGY R & D PROGRAMME Objective : Energy conservation	Project or Sector: Recycling of materials

Title:

..Conservation of energy and raw material due to the recycling of waste plastics extracted from urban garbage.

Duration: 30 months Period: 1/6/1977 - 30/6/1979	Contract N° : 202-76-7 EEB Project N° :

Contractor: C.R.I.F.

Address: 21 rue des Drapiers, 1050 Bruxelles

Head of Project: Mr. G. Viatour

Description of research work:

1. Objectives (aims)

Compared to recovery methods such as combustion with recuperation of energy, pyrolysis with recuperation of chemical products, reuse as charging materials, etc ..., the recovery methods by plastificagion are the only ones which make use of the inherent qualities of thermoplastic materials which constitute all their value. As a matter of fact it is not logical to construct a system of elimination of synthetic macromolecules, which is based on their decomposition or their combustion, whereas science has succeeded with much difficulty and energy spending to synthesize and stabilize them and to render them uncorruptible. Moreover, given the fact that thermoplastic materials can be melted and resolidified several times, it was normal that recovery methods by plastification gain an advantage over the others.

A complementary economic portion to the research effort will indicage the economic interest of plastics recycling on the EEC level and in particular the energy conservation possible through the realization of the proposed processes.

Total cost: FB. 15,983,517 u.a. 319,670	E.E.C. Contribution: 50 % FB. 7,991,758 u.a. 159,835

COMMISSION OF THE EUROPEAN COMMUNITIES	Project or Sector:
ENERGY R & D PROGRAMME	Recycling of materials
Objective : Energy conservation	

2. Work Programme

Phase 1
Economic_part :
- detailed study of the economic interest of recycling waste of plastic materials, taking mainly into account the potential market of recycled products on the level of the European Community;
- analysis of the technological and administrative conditions of the Member countries of the Community, influencing the penetration of recycling techniques for plastic materials contained in urban garbage.

Research_part :
- study of rheological and technical problems of the establishment and treatment of polymer alloys with industrial devices, on the basis of urban garbage;
- orientation study for a pilot installation capable of treating 600 t/y of plastic from urban garbage.

Phase 2

Realisation and operation of this pilot installation (not included in this contract).

3. Status

Work in progress.

Sector f:

Production of energy from waste

COMMISSION OF THE EUROPEAN COMMUNITIES ENERGY R & D PROGRAMME **Objective:** Energy conservation	**Project or Sector:** Production of energy from waste

Title:

    Fluidized bed combustion of low grade materials.

Duration: 15 months **Period:** 1/7/1977 - 1/10/1978	**Contract No:** 120-77 EEUK **Project No:**

Contractor: National Coal Board

Address: London SW1X 7AE, GB

Head of Project: Dr. J. Gibson

Description of research work

I. Objectives (aims)

The objective of this research programme is to investigate the best way of achieving combustion of waste material of very low calorific value in fluidized beds so as to maximise the recovery of useful heat from the system. Fluidized bed combustion is a new technique for burning material which offers advantages in heat transfer characteristics, flexibility, compactness and efficiency over conventional furnaces. This method would also enable low grade materials such as domestic refuse and unburnt colliery spoil to be burnt. The production of coal refuse alone can be estimated at around 150 million tons per year. The use of this refuse gives approximately 12 million tons per year of coal equivalent.

2. Work Programme

The programme consists of a design study of a unit of 5MN thermal capacity plus an experimental section on one or more pilot scale fluidized bed combustors.

The experimental section generates the inputs to evaluate the results of the design study and eventually produces the required detailed information.

The research programme has been integrated with complementary proposals by CERCHAR (E/F/020/F) and S.A. Heurtey (E/F/037/F). Agreement has been reached between these bodies and the National Coal Board to integrate and compare the results of the three programmes.

3. Status   Work in progress

Total cost:
    £   44,724
    u.a.   107,338
	E.C. Contribution: 50%
	    £ 22,362
	    u.a. 53,669

COMMISSION OF THE EUROPEAN COMMUNITIES ENERGY R & D PROGRAMME Objective : Energy conservation	Project or Sector: Production of energy from waste

Title:

Experimental study of the combustion of combustable matter having low calorific power, in order to determine its energy value.

Duration: 18 months Period: 14/6/1977 - 14/12/1978	Contract N° : 199-77 EEF Project N° :

Contractor: CERCHAR

Address: 33 rue de la Baume, F - 75008 Paris (France)

Head of Project: Mr. Busso

Description of research work:

1. Objectives (aims)

The essential purpose of this study is to permit the recovery of the energy contained in low grade combustible matter by pushing back the technical and economic boundaries which are at present accepted in the practice of combustion. Consequently, the following goals should be attained in succession :

1°   the establishment of a laboratory method for determining the thermal value in use of low grade combustible matter (such as bituminous shales, coal shales, etc ...)

2°   the choice of industrial techniques best adapted to the combustion of these substances. Tests carried out by means of pilot combustion furnaces will show up the difficulties and make easier the search for technical methods to reduce or overcome these difficulties;

3°   the evaluation of the cost of the recoverable energy according to an economic energy balance-sheet, drawn up at the time of these tests in pilot furnaces.

The expected recovery of energy which would result from a better use of natural resources, would increase the energy autonomy of these countries and of own, to a certain extent. It would permit the productive countries to supply rich fuel to the non-productive members of the Community.

Total cost: FF. 962,500 u.a. 173,299	E.E.C. Contribution: 50 % FF. 481,250 u.a. 86,650

COMMISSION OF THE EUROPEAN COMMUNITIES ENERGY R & D PROGRAMME Objective : Energy conservation	**Project or Sector:** Production of energy from waste

2. Work Programme

The suggested programme will therefore consist of three stages for each kind of fuel chosen :

First stage : study of the physico-chemical properties and characteristics of the fuel

The study will concentrate on six combustible materials :

- a Toarcian bituminous shale

- three coal shales, each having a different kind of mineral matrix : limestone, silico-aluminate, pyrites

- a ligno-cellulose waste (straw or wood)

- an urban waste treated as a replacement fuel.

In order to co-ordinate our studies with those proposed by the Heurtey Company and by the National Coal Board, we shall use batches of material from the same source, as far as this be possible.

Second stage : study of the combustion in an experimental furnace

Third stage : price estimate of industrial unit.

3. Status

Work in progress.

COMMISSION OF THE EUROPEAN COMMUNITIES ENERGY R & D PROGRAMME Objective : Energy conservagion	Project or Sector: Production of energy from waste

Title:

Fluidized bed gasification at low temperature of low grade materials.

Duration: 20 months Period: 7/6/1977 - 7/2/1979	Contract N° 200-77 EEF Project N° :

Contractor: S.A. Heurtey

Address: 30 - 32 rue Guersant, F - 75017 Paris

Head of Project: Mr. Michel Tamalet

Description of research work:

1. Objectives (aims)

The solid combustible materials of low caloric value are classified as waste materials and not considered as useful fuels for the existing boilers.

There are large quantities of these materials in the EC and they form a potential source of energy.

The annual production of coal refuse, residues from coal extraction can be estimated at around 150 millions tons per year, for the European Community.

The use of this refuse could give approximately the equivalent of 12 millions tons, per year, of coal equivalent.

The fluidized bed combustion permits the combustion temperature control, to avoid the melting of ashes, to suppress the cost of fine crushing which is even more interesting than the combustible is ashy.

The research programme will be integrated with complementary projects by Cerchar and National Coal Board and agreement has been reached between these these bodies and S.A. Heurtey to co-ordinate the three programmes by a series of regular meetings.

This research is designed to develop a process and equipment in order to valorize low grade materials by means of combustion in two stages :

- gasification by partial oxidizing at low temperature in fluidized bed

- post-combustion in a boiler or in a furnace.

Total cost: FF. 642,500 u.a. 115,682	E.E.C. Contribution: 50 % FF. 321,250 u.a. 57,841

COMMISSION OF THE EUROPEAN COMMUNITIES ENERGY R & D PROGRAMME Objective : Energy conservation	Project or Sector: Production of energy from waste

2.   Work Programme

- the study and design of pilot plant
- construction and perfecting the pilot plant
  - nature of low grade material : coal refuse
- measures of - organic material content
  - gazes composition after gasification
  - temperature of the flame
  - gazes composition after post combustion
  - ashes composition

3.   Status

Work in progress.

Sector g:

Evaluation of the specific energy consumption of equipment, processes and techniques

COMMISSION OF THE EUROPEAN COMMUNITIES ENERGY R & D PROGRAMME Objective: Energy conservation	Project or Sector: Assessment of the specific energy consumption of various equipment, processes and techniques

Title:

Optimized plant design to ensure energy conservation.

Duration: 18 months Period: 17/12/76 - 11/6/78	Contract No: 124-76 EEF Project No:

Contractor: Institut Français du Pétrole

Address: 1-4 Avenue du Bois-Préau, 92500 Rueil Malmaison

Head of Project: Mr. Y. Durandet

Description of research work

I. Objectives (aims)

Optimization of a plant, encompassing both process unit and utilities. This optimization consists of minimizing the total energy consumption, taking into account the economic conditions in relation to investments.

2. Work Programme

Consists of a study of the various devices leading to energy saving in a plant and in improving the method of selecting these devices and determining their optimum arrangement.

The following points are included:

- optimization method;

- study of various devices;

- study of actual cases, and in particular, a refinery.

3. Status

Work in progress.

Total cost: FF. 1,370,128 u.a. 246,683	E.C. Contribution: 55% FF. 753,570 u.a. 135,676

COMMISSION OF THE EUROPEAN COMMUNITIES ENERGY R & D PROGRAMME **Objective:** Energy conservation	**Project or Sector:** Assessment of the specific energy consumption of various equipment, processes and techniques

Title:

Conceptional study of steam turbine with multi and independently variable stage extraction rates of steam.

Duration: 18 months **Period:** 17/12/76 - 10/6/78	**Contract No:** 174-76 EEB **Project No:**

Contractor: Université Catholique de Louvain, Faculté des Sciences appliquées - Département thermodynamique et Turbomachines

Address: Bâtiment Stévin, 2, place du Levant
1348, Louvain-la-Neuve, Belgium

Head of Project: Prof. P. Wauters

Description of research work

I. Objectives (aims)

The aim of the contract is to execute a feasibility study and a performance evaluation of steam turbines which combine the generation of power with the extraction of large quantities of steam and showing an optimal performance under working conditions. This study also includes experimental work, as far as is required.

2. Work Programme

- Analysis of the problems and their precise technical formulation;

- Influence of the extraction of steam under steady and unsteady conditions;

- Possibilities for improving the steam extraction rate of an existing plan;

- Elaboration of complementary programmes for some of the calculations mentioned above;

- Compilation of some available experimental results and an attempted comparison with theoretical results;

- Overall feasibility study and final analysis and report.

3. Status
Work in progress.

Total cost: FB. 2,150,000 u.a. 43,000	E.C. Contribution: 100% FB. 2,150,000 u.a. 43,000

COMMISSION OF THE EUROPEAN COMMUNITIES ENERGY R & D PROGRAMME Objective : Energy conservation	Project or Sector: Assessment of the specific energy consumption of various equipment, processess and techniques.

Title:

To establish comparative energy consumption and efficiencies of different types of oil, gas and electric domestic space and hot water heating equipment on an annual basis.

Duration: 23 months Period: 14/6/1977 - 14/5/1979	Contract N° : 187-77 EEUK

Contractor: B.S.R.I.A.
Building Services Research and Information Association
Address: Old Bracknell Lane
Bracknell   GB - Berkshire RG12 4AH
Head of Project: Mr. P.A. Coles

Description of research work:

1. Objectives (aims)

The object of this research programme is to determine the comparative energy consumption of different types of appliances used for domestic space and water heating. The results of this investigation will be used to produce design guidance for the selection of the type of appliance that will result in the least energy consumption for a given application.

2. Work programme

The following stages are foreseen :
  stage 1. Pre trial preparation

- selection of appliances and of load profiles
- specification of control of system and of programmes
- design of test rig.

  stage 2. Construction of test rig

- purchase of all necessary equipment and appliances
- building and commissioning of test rig.

  stage 3. Laboratory testing

  stage 4. Analysis of data and preparation of final report

3. Status

Work in progress.

Total cost: £   51,792
               u.c.   124,301
	E.E.C. Contribution: 50 %
	               £ 25,896
	               u.c. 62,150

COMMISSION OF THE EUROPEAN COMMUNITIES ENERGY R & D PROGRAMME Objective: Energy conservation	Project or Sector: Assessment of the specific energy consumption of various equipment, processes and techniques

Title:

Energy consumption in electric water heating dependent on the technology of the apparatus and the pattern of demand.

Duration: 15 months Period: 1/2/1977 - 1/5/1978	Contract No: 190-77 EED Project No:

Contractor: Gesellschaft für Praktische Energikunde G.F.P.E.

Address: Am Blütenanger 71, D-8000 München

Head of Project: Prof. Dr. Ing. H. Schaefer

Description of research work

  I. Objectives (aims)

  Energy consumption for water heating is the second largest demand category in the overall end-consumption in households, coming after space heating.   A study is to be carried out to obtain a general picture of the influence of apparatus design in energy consumption for water heating; the study will cover both generating and consuming equipment.   The purpose of the research is to determine and compare the operational behaviour and specific consumption of various types of appliance.   The information so obtained will be used for decisions in the field of energy and particularly for formulating energy-conservation measures.

  2. Work Programme

  Determination and comparison of operational behaviour and specific energy consumption of various types of appliance, to obtain a basis for decisions in the energy field, and particularly for formulating energy-conservation measures.

  The programme is composed of two sub-programmes :

  (a)   Technology of electric water heating

  (b)   Techniques of temperature mixing.

  3. Status

  Work in progress

Total cost:
    DM.   258,935
    u.a.   70,747
	E.C. Contribution: 100%
	    DM. 258,935
	           70,747

COMMISSION OF THE EUROPEAN COMMUNITIES ENERGY R & D PROGRAMME Objective : Energy conservation	Project or Sector: Assessment of the specific energy consumption of various equipment, processess and techniques

Title:

Development and promotion of methods for reduction of energy consumption during farm cooling of milk.

Duration: 26 months	Contract N° : 215-76 EEN
Period: 1/1/1977 - 1/3/1979	Project N° :

Contractor: Nederlands Institut voor Zuivelonderzoek

Address: Kernhemseweg 2, Ede, The Netherlands

Head of Project: Ir. J. Ubbels

Description of research work:

1. Objectives (aims)

The introduction of tank-cars for the collection of milk on the farms and the transport to the dairies every two or three days faced the farmers with the need to cool the milk and store it in thanks from 35° C to below 4° C, which corresponds with a temperature fall 31° K, is effected within 3 h by an electrically driven refrigeration unit of which the evaporator is usually formed by the bottom of the farm tank. At present about 40 % of the milk produced in the Netherlands is cooled at the farms. It is expected that in 1980 the corresponding percentage will amount to 100.
This refrigeration consume 5 to 6 times the amount of electrical energy needed at the dairy farm.

The objective of the project is to present reliable methods and equipment to reduce this consumption of energy. This objective will be worked out along two lines :

- before the milk enters the refrigerated farm tank it is cooled from 35° C to a temperature between 15 and 20° C in a precooler in which cold water (10-to 15° C) flows in counter current with the milk. The water leaving the precooler is reused for example as drinking water for cows.

- the heat of condensation which comes free in the condensor of the refrigeration unit is used to preheat water to temperatures between 40 and 50° C (flow sheet 3). Sometimes this water is used directly, mostly it is fed to a boiler were it is heated additionally to 85° C.

Total cost:
    Fl. 404,315
    u.a. 111,689
	E.E.C. Contribution: 50 %
	    Fl. 202,158
	    u.a. 55,844

COMMISSION OF THE EUROPEAN COMMUNITIES ENERGY R & D PROGRAMME Objective : Energy conservation	**Project or Sector:** Assessment of the specific energy consumption of various equipement, processess and techniques.

2.   Work Programme

Phase 1

- development of the precooler
- construction ·
- installing and testing of the precooler
- assessment of the precooler; the results of the testing will be assessed in relation to the optimal cooler for the different farm sized and the implementation of the precooler (technico economical aspects)

Phase 2

Implementation of the precooler

- construction of 10 precooler
- installing of precooler in farms
- collecting of information of the behavious of the precooler under practical condition

Phase 3

Production of hot water

- need of hot water ·
- data aquisition, tests and assessment.

3.   Status

Work in progress.

Sector h:

Development of methods for storage
of secondary energy

COMMISSION OF THE EUROPEAN COMMUNITIES ENERGY R & D PROGRAMME Objective: Energy conservation	Project or Sector: Development of methods of accumulating secondary energy

Title:

Optimization of a process of shaping of alumina beta powder for Na.S. batteries.

Duration: 24 months Period: 1/1/1976 - 1/1/1978	Contract No: 179-77 EEF Project No:

Contractor: Laboratoires de Marcoussis, Centre de recherches de la compagnie générale d'électricité

Address: Route de Nozay - 91460 Marcoussis - France

Head of Project: Mr. Vie

Description of research work

I. Objectives (aims)

The aim of this research is to solve the economic and technical problems posed by the development and industrial manufacture of an electrolyte of alumina beta required in the building of sodium sulphur (NA.S) generators for the storage of electrical energy, particularly in order to regulate supply to the grid.

This application requires that, as well as having high performance characteristics (low resistivity, advanced physico-chemical and electro-chemical properties, a life expectancy of over ten years) the solid electrolyte be manufactured at particularly low cost and by an extremely reliable process.

The contractor is engaged in more extensive studies, outside the scope of this contract, into the development of the other components of the accumulator.

2. Work Programme

The programme is divided into two stages, the contract only covers the first stage.

Stage 1: - The development of a process of synthesis of alumina β powder;

- Development of a process of shaping by electrophoresis before moving over to industrial production;

../..

Total cost: FF. 1,751,393 h.t. u.a. 315,330	E.C. Contribution: 50% FF. 1,050,836 t.t.c. u.a. 189,203

COMMISSION OF THE EUROPEAN COMMUNITIES ENERGY R & D PROGRAMME Objective: Energy conservation	Project or Sector: Development of methods of accumulating secondary energy

2. Work Programme cont...

- Study of a conventional alumina β sintering process at low temperatures.

3. Status

Work in progress

PROPOSALS

UNDER NEGOTIATION

PROPOSALS UNDER NEGOTIATION

Sector a: Improved insulation of buildings

1. Optimization of multi-functional insulating glazings.

    Flachglas AG, Delog-Detag, Germany

2. Preliminary normalization study of typical thermal insolation
    failures by infrared thermography.

    Laboratoire National d'Essais, France

Sector b: Use of heat pump

1. The development of a thermally driven heat pump for the heating
    of a single family house, with particular reference to the
    operation at relatively low temperature, in order to exploit
    different heat sources.

    Zanussi, Italy

2. Development of commercially viable air-source heat pumps for
    domestic space and water heating in Northern European climatic
    conditions.

    University of Ulster, Galway and Coleraine, Ireland

Sector c: Urban Transport

1. Potential for energy conservation by a shift to other types of
    engines in passenger cars.

    T.N.O., Netherlands

Sector d: Residual heat recovery

1. Technical and economical possibilities for residual heat
    recovery in the aluminium industry.

    Vereinigte Aluminium Werke, Germany

2. Residual heat recovery in buildings and power plants.

    Battelle Frankfurt, Germany

3. Development of a method for heat recovery in the coke industry by cooling coke and the testing of such an installation.

    Stahlwerke Röchling Burbach, Germany

4. Residual heat recovery by combining power supply and LNG evaporation in a closed gas turbine.

    Gutehoffnungshütte Sterkrade AG, G. H. H., Germany

5. Energy recovery in ferrous scrap melting plants.

    Universita di Genova, Italy

Sector e: Materials recycling

1. Survey of the possibilities for reuse of thermoset scrap from industry in materials for building and public works.

    IRCHA, Institut National de Recherche Chimique Appliquée, France

Sector g: Evaluation of the specific energy consumption of equipment, processes and techniques

1. To evaluate the methodology and practice of boiler efficiency testing with the object of identifying superior measurement techniques which may be applied to improve energy conservation.

    Institute for Industrial Research, Ireland and Denmark

2. The calculation of the specific energy consumption for industrial processes by reference to the first and second law of thermodynamics.

    Battelle Institut, Frankfurt, Germany

3. Financial and economic parameters of energy saving investments in the French chemical industry.

    Institut National Polytechnique de Nancy, France

4. The optimum arrangement of multi-boiler installation, with regard to power distribution in order to obtain a high average yearly efficiency.

    Gesellschaft zur Förderung der Heizungs- und Klimatechnik M. B. H., Germany

5. Economy and energy saving in the textile industry.

    Institut Textile de France, France

6. Engineering research study on oxygen combustion technology.

    Research and Development Company of Ireland Ltd., Ireland

7. Energy consumption and energy saving in the heat setting and
    dye fixation of textile fabrics on stenters.

    Shirley Institute, United Kingdom

8. Utilization and conservation of energy in the European Food
    Industry.

    The British Food Manufacturing Industries Research
    Association, United Kingdom

9. Energy saving in paper making by means of increased water
    removal.

    T.N.O., Netherlands

Sector h: Development of methods for storage of secondary
energy

1. Research and development of systems for thermal energy storage
    in a temperature range of -25 to 150°C.

    Philips, Germany

2. Galvanic high energy batteries with melt electrolytis.

    Varta Batterie, Germany

Production and utilization of hydrogen

Hydrogen is an interesting energy storage and transportation medium.
As a fuel it causes no ost-combustion pollution.   It can be produced
from water by electrolysis, stored and distributed in liquid or gaseous
form or combined with a metal, reconverted into electricity in a fuel
cell, burned to produce heat, transformed into mechanical power in a
conventional internal combustion engine or used as a raw material in
the chemical, petrochemical and metallurgical industries.

The potential market is considerable, because of the increasing trend
towards the use of gaseous fuels and of electricity.

If it were shown to be possible to obtain hydrogen under economically
acceptable conditions by electrolysis or by thermo-chemical dissociation,
using the nuclear heat of the high-temperature reactors, this might
provide at long range an alternative to electricity and maybe replace
some liquid or gaseous fuels.

The present programme contains two predominant parts concerned with
hydrogen production and a third part devoting some effort to the study
of problems related to this utilization of hydrogen.   This distribution
of efforts is based on the philosophy that one should first find
improved methods of producing cheaper hydrogen from water before
generating new information on its possible use.

Basically, the programme concentrates on the following projects:

A.   Thermochemical production of hydrogen

This chapter deals with both purely thermochemical cycles and hybrid
cycles appropriated to produce hydrogen from water.   In these cycles,
adjuvant reactants are continuously recycled.

In the contracts concluded up to now (first phase), the work mainly
consists of the determination of chemical, thermodynamic and kinetic
data on promising reactions, and in studies on the separation of the
products of single reactions and the economic evaluations of whole
cycles.   For cycles which prove to be promising, bench scale experiments
could be built to demonstrate their feasibility.   Up to now, no contract
has been placed for this purpose.

B.   Electrolytic production of hydrogen

Alcaline electrolysis consists in electrolyzing an aqueous solution of
KOH (usually with a potash content of 30%) between two (metal) electrodes
at atmospheric or higher pressure.   In the first case, the electrolysis
temperature is close to $80°C$.

The objective of the research programme is to determine optimized
electrolysis conditions (temperatue, pressure, current density, etc.)
in which the energy consumption would be minimum and the hydrogen
production cost as low as possible.

Work is carried out in particular on electrodes in order to reduce the
over voltage and on diaphragm materials with improved stability at
higher temperatures i.e., 150°C and over.

The programme also covers the solid polymer electrolyte process which
has been shown to hold great promise, although using rather expensive
materials.   Alternative cheaper materials with adequate performances
have to be found (electrocatalysts, structures and electrolyte) before
the process becomes competitive.

High temperature electrolysis is processing water vapor at temperatures
between 800 and 1000°C; the electrolyte is a solid membrane of an
oxygen-ion conducting ceramic materials, in particular stabilized
zircona.   Experimental work under way is intended to develop adequate
electrolyte materials and suitable electrodes.   It should also
establish the feasibility and viability of the process.

For those processes which will have reached sufficient maturity, prototypes could be built in a later phase for long life testing.

C. Utilization of hydrogen

Much information already exists on the potential uses of hydrogen.   The
programme carries out studies dealing with applications not yet well
covered, (steel and coal industry, etc.).

The safety aspects of hydrogen production and uses are considered in
detail in order to clarify which hazards are involved.

Transportation in pipelines is investigated and some effort is devoted
to technical problems arising from safety and economic requirements.
Small scale storage of hydrogen (cryogenic and hydrides) is also
included.

SUMMARY AND BREAKDOWN
OF FUNDING

Objective: Production and utilization of hydrogen

Project	Number of contracts(*)	Total cost u.a.	E. C. contribution u.a.	Number of proposals under negotiation
A	6	564,433	328,197	-
B	14	2,204,612	1,159,780	-
C	9	624,593	340,258	-
Total	29	3,393,638	1,828,235	-

(*)Signed both by the Commission and the Contractor or sent for
signature to the Contractor.

Project A

Thermochemical production of hydrogen

COMMISSION OF THE EUROPEAN COMMUNITIES ENERGY R & D PROGRAMME **Objective:** Production and utilisation of hydrogen	**Project or Sector:** Thermochemical production of hydrogen

Title:
    Automatic flowsheet synthesis and optimization.

Duration: 12 months **Period:** 1/12/76 - 30/11/77	**Contract No:** 038-76 EHI **Project No:** H/A9(I)

Contractor: Analysis and Development of Energy Systems S.R.L. (ADES)

Address: Via della Trinità dei Pellegrini, 19, I-00126 Roma

Head of Project: Prof. C. Mustacchi

Description of research work

I. Objectives (aims)

Definition and approximate evaluation of flowsheets of thermochemical processes for water splitting.

2. Work Programme

All available information on the cycles studied will be collected (thermodynamic data etc.). Data which is not available will be obtained by extrapolation and approximation.

On the basis of a starting flowsheet and of thermal and mass balances an optimisation of the plant will be done.

The output of this process will supply all the relevant information: flowsheets, detailed mass and thermal balances, apparatus dimensions, evaluation of the cost of various pieces of equipment, of the operative costs and the influence of the main parameters on these evaluations.

3. Status

Just started.

Total cost:
     Lit   24,700,000
     u.a.   39,520
	E.C. Contribution: 60%
	     Lit 14,820,000
	     u.a. 23,712

COMMISSION OF THE EUROPEAN COMMUNITIES ENERGY R & D PROGRAMME Objective: Production and utilization of hydrogen	Project or Sector: Thermochemical production of hydrogen

Title:

Hybrid processes for hydrogen production.

Duration: 12 months Period: 4/11/76 - 3/11/77	Contract No: 056-76 EHD Project No: H/A5(D)

Contractor: Deutsche Forschungs- und Versuchsanstalt für Luft- und Raumfahrt e.V. (DFVLR)

Address: Pfaffenwaldring 38-40, D-7000 Stuttgart 80

Head of Project: Prof. W. Peschka

Description of research work

I. Objectives (aims)

To search for and evaluate possible electrolytic reactions to be coupled with thermochemical reactions to form cycles.

2. Work Programme

The work is concentrated mainly on the study of the electrolytic step. The main criteria for cycle evaluation are: a maximum temperature lower than $950^{\circ}C$, limited number of reactions, limited cost and high efficiency.

The electrolytic step will be chosen on the basis of: low electrical energy requirements, high current density, low overvoltage and simple operation.

Suitable reactions for cycle closure will be looked at for the most promising electrolytic steps and evaluation of the cycles will be done.

3. Status

The collection of basic data has commenced.

Total cost:
 	
    DM   364,900
    u.a.   99,700
	E.C. Contribution: 50%
	
	    DM 182,450
	    u.a. 49,850

COMMISSION OF THE EUROPEAN COMMUNITIES **ENERGY R & D PROGRAMME** **Objective:** Production and utilization of hydrogen	**Project or Sector:** Thermochemical production of hydrogen

Title:

Efficiency of water dissociation to time–resolved Hydrogen and Oxygen products through non–equilibrium processes on catalyst surfaces.

Duration: 12 months **Period:** 4/11/76 – 3/11/77	**Contract No:** 059–76 EHEIR **Project No:** H/A1(EIR)

Contractor: University College, Cork

Address: Cork, Ireland

Head of Project: Prof. Cunningham

Description of research work

I. Objectives (aims)

To investigate the release of hydrogen and oxygen by the dissociation of water under pulsed electrocatalytic procedures.

2. Work Programme

Semiconductor catalysts in thin layers, such as titanium dioxide, zinc oxide, nickel monoxide, iron oxides and elemental semiconductors.

The systems showing electrocatalytic activity will be examined at different temperatures and subjected to various electrical conditions to check the overall efficiency and possible practical interest of the proposed process.

3. Status

Just started.

Total cost:	**E. C. Contribution:** 63 %
£ 14,900 u.a. 35.760	£ 9,400 u.a. 22.560

COMMISSION OF THE EUROPEAN COMMUNITIES **ENERGY R & D PROGRAMME** **Objective:** Production and utilization of hydrogen	**Project or Sector:** Thermochemical production of hydrogen

Title:

Study of a family of thermochemical cycles, based on carbon oxides, metallic oxides and carbonates.

Duration: 12 months **Period:** 3/12/76 - 2/12/77	**Contract No:** 066-76 EHF **Project No:** H/A3(F)

Contractor: Commissariat à l'Energie Atomique (CEA SACLAY)

Address: F-91190 Gif-sur-Yvette

Head of Project: Mr. P. Courvoisier (Centre d'Etudes Nucléaires de SACLAY)

Description of research work

I. Objectives (aims)

The possible thermochemical cycles based on oxides of carbon, metallic carbonates and oxides will be explored. The thermodynamic data on the metallic carbonates will be evaluated.

2. Work Programme

The thermal decomposition of metallic carbonates will be studied, as well as their formation (from CO_2 and oxides), under various conditions.

The thermodynamic data on all reactions will be collected from literature or measurements taken where necessary.

More detailed measurements will be taken of promising reactions; the operative conditions of possible cycles, the tentative flowsheet and material balances will be established also.

3. Status

The collection of thermodynamic data has been started.

Total cost:	**E. C. Contribution:** 50%
FF 621,480 u.a. 111,898	FF 310,740 u.a. 55,949

COMMISSION OF THE EUROPEAN COMMUNITIES ENERGY R & D PROGRAMME **Objective:** Production and utilization of hydrogen	**Project or Sector:** Thermochemical production of hydrogen

Title:
The catalysis of the chemical and electrochemical transformations of compounds (SO_2 etc.) within the framework of the hybrid processes for hydrogen production from water.

Duration: 12 months **Period:** 10/5/77 - 9/5/78	**Contract No:** 162-76 EHD **Project No:** H/A8(D)

Contractor: Kernforschungsanlage Jülich GmbH

Address: Postfach 1913, D-517 Jülich 1

Head of Project: Dr. H. Barnert

Description of research work

I. Objectives (aims)

To evaluate the parameters for the economic feasibility of the electrolytic oxidation of SO_2 to H_2SO_4.

2. Work Programme

High performance anodes of non-precious materials will be looked at. After this screening, tests in H_2SO_4, at 60% wt. and at 80°C will be undertaken. Optimal operating conditions will be looked for in connection with the cycle optimisation.

Exploratory work on homogeneous catalysts (I/I^-, Me^{z+}/Me^{z+1}) will be done, preceded by the study of possible methods of catalyst separation; conditions should be investigated which are compatible with the economics of the complete cycle.

For I/I^-, initially, the conditions of stability (not leading to by-products) will be determined, as well as the influences of anode materials and SO_2 pressure.

3. Status

Just started.

Total cost: DM. 593,440 u.a. 162,142	**E.C. Contribution:** 50% DM. 296,980 u.a. 81,142

COMMISSION OF THE EUROPEAN COMMUNITIES ENERGY R & D PROGRAMME **Objective:** Production and utilization of hydrogen	**Project or Sector:** Thermochemical production of hydrogen

Title:

Separation processes in thermochemical hydrogen production.

Duration: 12 months **Period:** 1.2.77 - 31.1.78	**Contract No:** 185-77 EHD **Project No:** H/A4(D)

Contractor: Rheinisch-Westfälische Technische Hochschule Aachen

Address: Templergraben 55, D-5100 Aachen

Head of Project: Prof. Dr. Ing. K.F. Knoche

Description of research work

I. Objectives (aims)

Within the framework of the thermochemical cycles of the sulphur family, codes for the design of absorption columns, stripping and distillation columns, will be developed. Phase equilibria of the H_2O/H_2SO_4 system will be measured, along with the gas-liquid mass transfer of SO_2-O_2 in H_2O.

2. Work Programme

Computer codes will be developed for the design of adiabatic/isothermal absorption columns with intermediate feeds and spills, the design of distillation and stripping columns with feeds and spills and the automatic extrapolation of existing thermodynamic data. The codes will be used to check the economic and practical feasibility of the separation processes foreseen for the family of sulphur cycles under development.

The phase equilibria of H_2O/H_2SO_4 will be experimentally determined up to $500^\circ C$ and 100% H_2SO_4 at pressures up to 15 bars.

The kinetics of the SO_2+O_2 absorption in water (for the design of absorption columns) will be experimentally determined at temperatures up to $50^\circ C$, pressures up to 50 bars and up to 80% SO_2.

3. Status

Just started.

Total cost: DM. 422,410 u.a. 115,413	**E.C. Contribution:** 82% DM. 347,643 u.a. 94,984

Project B

Electrolytic production of hydrogen

COMMISSION OF THE EUROPEAN COMMUNITIES ENERGY R & D PROGRAMME Objective: Production and utilization of hydrogen	Project or Sector: Electrolytic production of hydrogen

Title:
Anodic materials for water electrolysis.

Duration: 12 months Period: 3/12/76 - 2/12/77	Contract No: 039-76 EHI Project No: H/B21(I)

Contractor: Università di Milano, Istituto di Elettrochimica e Metallurgia

Address: Via Festa del Perdono, 7, I-20122 Milano

Head of Project: Prof. G. Fiori

Description of research work

I. Objectives (aims)

To study the properties of mixed oxides as anode catalysts in
electrolytic cells working with KOH solutions.

2. Work Programme

A series of mixed oxides containing one oxide of a transition metal,
and one oxide of one of the lanthanides will be prepared.

The basic properties of thin films of the mixed oxides will be
determined, such as: the crystal structure, the electrical conductivity,
the current versus voltage in electrolysis, the kinetics of the oxygen
discharge. Some tests will be done on the long term stability in KOH
solutions at moderate temperatures.

The possibility of supporting the mixed oxides by one of the constituent
metals will be evaluated.

3. Status

Preparation and evaluation of the mixed oxides has started.

Total cost:
    Lit. 47,690,000
    u.a. 76,304
	E. C. Contribution: 56%
	    Lit. 26,700,000
	    u.a. 42,720

COMMISSION OF THE EUROPEAN COMMUNITIES ENERGY R & D PROGRAMME Objective: Production and utilization of hydrogen	Project or Sector: Electrolytic production of hydrogen

Title:

High temperature electrolysis of water vapour on flat, solid, thin-film electrolyte.

Duration: 12 months Period: 1/10/76 - 30/9/77	Contract No: 040-76 EHF Project No: H/B3(F)

Contractor: Commissariat à l'Energie Atomique (C.E.N. Grenoble)

Address: rue de la Fédération, 29 - 75015 Paris

Head of Project: Mr. J. C. Blin

Description of research work

I. Objectives (aims)

To develop a half cathodic cell in thin-film plane geometry for high temperature vapour electrolysis.

2. Work Programme

Preliminary studies of the water vapour reduction will be done as a function of hydrogen content, temperature and flow-rate. Permeability of electrolytes to hydrogen and their conductivity will be measured.

Yttrium stabilized zirconia electrolytes will be fabricated by various techniques; the influence of the fabrication methods on the properties and the performances obtained will be correlated.

Permeability, fissure development, resistance to thermal cycling and electrical parameters will be measured.

The satisfactory electrolytes obtained will be tested in cells operating under practical conditions.

3. Status

Part of the preliminary work has been done, material development and fabrication technologies are proceeding.

Total cost: FF. 1,455,000 u.a. 261,973	E.C. Contribution: 50% FF. 727,500 u.a. 130,987

COMMISSION OF THE EUROPEAN COMMUNITIES ENERGY R & D PROGRAMME **Objective:** Production and utilization of hydrogen	**Project or Sector:** Electrolytic production of hydrogen

Title:

High temperature electrolysis with ZrO_2 solid electrolytes.

Duration: 12 months **Period:** 13/9/76 - 12/9/77	**Contract No:** 045-76 EHD **Project No:** H/B6(D)

Contractor: Brown, Boveri & Cie. AG., Mannheim

Address: Zentrales Forschungslabor, Eppelheimer Str. 82, D-6900 Heidelberg

Head of Project: Dr. F. J. Rohr

Description of research work

I. Objectives (aims)

To characterize high temperature electrolytic cells of tubular form at temperatures around $1,000^{\circ}C$.

2. Work Programme

The electrolytic cell will be characterized: the voltage/current characteristic will be determined as a function of H_2/H_2O ratio, of the temperature ($700-1,000^{\circ}C$), of the thickness of the electrolyte (0.3-1 mm.).

Measurements of long duration at $1,000^{\circ}C$ and 0.5-1 A/cm^2. with $H_2/H_2O \leq 0.2$, to check the stability of the electrolyte, of the electrode/electrolyte interface, the effect of overvoltage on the stability will be made.

Multicellular units (of at least 10 cells) will be tested for life and efficiency.

3. Status

Satisfactory operation of cells at 0.5 A/cm^2. and $1,000^{\circ}C$ has been obtained up to 1,000 hours. Characterization of the cells is progressing. A first three-cell module is in preparation.

Total cost: DM. 635,830 u.a. 173,724	**E.C. Contribution:** 50% D.M. 317,915 u.a. 86,862

COMMISSION OF THE EUROPEAN COMMUNITIES **ENERGY R & D PROGRAMME** **Objective:** Production and utilization of hydrogen	**Project or Sector:** Electrolytic production of hydrogen

Title:

Water electrolysis at high pressure (30-400 bars) and medium temperature (200-500°C).

Duration: 12 months **Period:** 13/9/76 - 12/9/77	**Contract No:** 047-76 EHD **Project No:** H/B1(D)

Contractor: Technische Hochschule Darmstadt

Address: Petersenstrasse 15, D-1600 Darmstadt

Head of Project: Prof. H. Wendt

Description of research work

I. Objectives (aims)

To assess the feasibility of a high pressure, medium temperature electrolytic process, its basic problems and its economics.

2. Work Programme

Small (100 cc.) autoclaves will be made for electrochemical measurements at medium temperature.

The vapour pressure of KOH (and NaOH) solutions will be measured. The cell voltage/current curves up to $2A/cm^2$ at 200-500°C and 30-400 bars will be measured, and also the specific conductivity of high concentrations of KOH as a function of its composition.

Longer term measurements, selection of suitable electrode materials, their aging behaviour, the potential of half-cells and their change with the main parameters, will be tested in more elaborate autoclaves.

Corrosion resistance of specific materials will be evaluated in the range 200-400°C for 200 hrs.

3. Status

Data collection from literature has been done, as well as some corrosion tests and apparatus construction.

Total cost: DM. 293,000 u.a. 80,056	**E.C. Contribution:** 75% DM. 220,000 u.a. 60,109

COMMISSION OF THE EUROPEAN COMMUNITIES ENERGY R & D PROGRAMME Objective: Production and utilization of hydrogen	Project or Sector: Electrolytic production of hydrogen

Title:

      Investigations on the improvement of asbestos diaphragms
      and their possible replacement by semipermeable membranes.

Duration: 12 months Period: 1/1/77 - 31/12/78	Contract No: 058-76 EHD Project No: H/B25(D)

Contractor: Krebskosmo Gesell. für Chemie-Ingenieur-Technik mbH.

Address: Zeltinger Platz 16, D-1000 Berlin 28

Head of Project: Dr. E. Hausmann

Description of research work

I. Objectives (aims)

To improve the performances of diaphragms of electrolytic cells.

2. Work Programme

The diaphragm materials procured from commercial producers or
specially made will be tested.   These materials will include:
straight asbestos paper; asbestos paper with organic polymer
impregnation; organic polymers, synthetic ceramic materials; "Nafion".

The evaluation will cover the following properties: porosity,
permeability, electrical resistivity, corrosion resistance (in KOH
concentrated at 80-90°C) in electrolytic cells.

Satisfactory materials will be tested for at least 1,000 hours in a
laboratory cell.   A test in a pilot plant or small scale unit will
be done on promising materials.

3. Status

Just started.

Total cost:
    DM.   192,774
    u.a.   52,670
	E.C. Contribution: 50%
	    DM. 96,387
	    u.a. 26,335

COMMISSION OF THE EUROPEAN COMMUNITIES ENERGY R & D PROGRAMME **Objective:** Production and utilization of hydrogen	**Project or Sector:** Electrolytic production of hydrogen

Title:
Development and parametric testing of water electrolysis cells for hydrogen production.

Duration: 12 months **Period:** 1/9/76 - 31/8/77	**Contract No:** 060-76 EHB **Project No:** H/B2(B)

Contractor: Centre d'Etudes de l'Energie Nucléaire (C.E.N.)

Address: Avenue E. Plasky 144 - 1040 Bruxelles

Head of Project: Dr. L. H. Baetsle

Description of research work

I. Objectives (aims)

To improve the cell technology aimed at obtaining cells working at up to 150°C, 20 bars with 1.65V across the cell at 10.000 A/m^2.

2. Work Programme

The electrodes will be similar to the miltilayer structure already developed by the contractor.

Optimiation will be done on this structure to obtain the highest performances possible, while limiting the cost. The electrolyte will be solid, chosen between: asbestos formulations; inorganic ion exchanges; organic ion exchangers.

The screening tests will be done on 4 cm^2 cells at 70°C. Larger cells of 36 cm^2 will be developed for operation at 20 bars and 150°C in autoclaves containing stacks of 5 cells, equipped for a complete characterization of the modules. A preliminary study of 1 KW modules will be done.

3. Status

Elementary cells have given promising results; testing facilities have been built for screening and further development; the design of the 36 cm^2 module has been initiated.

Total cost: FB 17,785,150 u.a. 355,703	**E.C. Contribution:** 50% FB 8,892,575 u.a. 177,851

COMMISSION OF THE EUROPEAN COMMUNITIES ENERGY R & D PROGRAMME **Objective:** Production and utilization of hydrogen	**Project or Sector:** Electrolytic production of hydrogen

Title: Electrolysis of water in an alkaline medium at higher temperatures and pressures.	

Duration: 12 months **Period:** 1/10/76 - 30/9/77	**Contract No:** 061-76 EHN **Project No:** H/B5(NL)

Contractor: Organization for Applied Scientific Research (TNO)

Address: Central Laboratory, P.O. Box 217, Delft (Netherlands)

Head of Project: Dr. Ch. A. Kruissink

Description of research work

I. Objectives (aims)

To acquire a general know-how on construction principles of electrolytic cells, in particular on the influence of the form and configuration of the electrodes.

2. Work Programme

Special half-cells will be used, with interchangable electrodes to study their influence on the performances. Current densities will range from 0.1 to 1 A/cm²: voltage drop, electrode potential, temperatures, electrolyte flow and the influence of gas bubbles will be measured.

The main parameters will be voltage, distance between electrode and diaphragm and electrolyte flowrate. Electrocatalytic surfaces (metals, alloys, oxides, carbides) will be evaluated for the anode at up to 150°C in KOH. The most promising catalysts will be tested for 1,000 hours at up to 1,500 A/cm².

Electrodes of high activity and a large specific surface area will be developed. The base material will be nickel plus the catalysts developed. Short screening tests will be done to choose the best electrodes to be submitted for 1,000 hour tests.

3. Status

Cell construction is complete, pulsed current measurements have been done. Tests for preparing large specific area electrodes have been done.

Total cost: Fl. 773,256 u.a. 213,607	**E.C. Contribution:** 50% Fl. 386,628 u.a. 106,803

COMMISSION OF THE EUROPEAN COMMUNITIES ENERGY R & D PROGRAMME **Objective:** Production and utilization of hydrogen	**Project or Sector:** Electrolytic production of hydrogen

Title:

   Amelioration of the electrolytic process for hydrogen production.

Duration: 12 months **Period:** 4/10/76 - 3/10/77	**Contract No:** 062-76 EHF **Project No:** H/B11(F)

Contractor: Centre de Recherche de la Compagnie Générale d'Electricité.

Address: Route de Nozay, F 91460 Marcoussis

Head of Project: Mr. Appleby

Description of research work

   I. Objectives (aims)

   To study the possibility of developing high efficiency diaphragms and electrodes for the electrolytic hydrogen production.

   2. Work Programme

   Teflon bonded potassium titanate and potassium zirconate diaphragms will be studied and optimized, they will be characterised and their potential cost evaluated.

   High performance electrodes of sintered nickel base, impregnated with various catalysts will be tested.   Screening electrolytic tests and stability tests in KOH at 150°C will be done.   The most promising electrodes will be characterised and tested up to 2,000 hours and up to 150°C.   Their potential cost will be evaluated.

   3. Status

   A potentially good diaphragm has been obtained;   the earlier tests on impregnated electrodes have also been done.

Total cost:
   FF.   923,280
   u.a. 166,237
	E.C. Contribution: 50%
	   FF. 461,640
	   u.a. 83,118

COMMISSION OF THE EUROPEAN COMMUNITIES ENERGY R & D PROGRAMME **Objective:** Production and utilization of hydrogen	**Project or Sector:** Electrolytic production of hydrogen

Title:

Thermodynamic study of the conditioning of hydrogen and oxygen produced by electrolysis at moderate temperatures.

Duration: 12 months **Period:** 4/10/76 - 3/10/77	**Contract No:** 063-76 EHF **Project No:** H/B15(F)

Contractor: C.E.M. Cie Electro-Mécanique, Département TA

Address: 37, Rue du Rocher, F-75008 Paris

Head of Project: Mr. Prost

Description of research work

I. Objectives (aims)

To recuperate water and the heat of compression from the effluence of an electrolytic cell.

2. Work Programme

The thermodynamic properties of water vapour and hydrogen mixtures will be determined for the operating conditions of the electrolyzer.

A compressor will be designed. The compression heat and condensation heat will be fed back to the electrolyzer.

The performances of such a system will be evaluated.

3. Status

The thermodynamic data for wet hydrogen has been determined, a first approximation of the compressor conceptual design has also been done.

Total cost:
    FF.   300,000
    u.a.   54,015
	E.C. Contribution: 50%
	    FF. 150.000
	    u.a. 27,008

COMMISSION OF THE EUROPEAN COMMUNITIES ENERGY R & D PROGRAMME **Objective:** Production and utilization of hydrogen	**Project or Sector:** Electrolytic production of hydrogen

Title:

Study of new anodic and cathodic materials and of new diaphragms for the electrolytic production of hydrogen.

Duration: 12 months **Period:** 1/1/77 - 31/12/77	**Contract No:** 067-76 EHI **Project No:** II/B23(I)

Contractor: Oronzio de Nora, Impianti Elettrochimici Spa,

Address: Via Bistolfi 35, I-20134 Milano

Head of Project: Dr. Spaziante

Description of research work

I. Objectives (aims)

To develop new electrodes and catalysts and new diaphragms for the electrolysis of water.

2. Work Programme

The electrochemical and corrosion behaviour of the basic materials for electrode construction will be examined in KOH 29% at between 70 and 100°C. Iron, nickel together with their alloys and compositions will be investigated.

The transition metals, with the addition of TiO_2, ZrO_2, sulphur etc. will be studied as possible catalysts, using various preparation techniques. Screening tests will be done to choose the most promising possibilities for further development. Longer tests will be done in KOH 29%, at current densities between 1 and 10 kA/m^2 eventually up to 130°C.

The best catalysts will be optimized and completely evaluated. The bulk of the work will be concentrated on the anodic catalysts, but some minor development of cathodic catalysis will be done.

Possible improvements to the asbestos diaphragms will be studied in order to increase their stability at temperatures above 80°C and their electrical conductivity.

../..

Total cost: Lit. 191,814,000 u.a. 306,902	**E.C. Contribution:** 50% Lit. 95,907,000 u.a. 153,451

COMMISSION OF THE EUROPEAN COMMUNITIES ENERGY R & D PROGRAMME **Objective:** Production and utilization of hydrogen	**Project or Sector:** Electrolytic production of hydrogen

2. Work Programme cont....

Their resistance to cycling between 90 and 130°C in 29% KOH solutions at a current density around $5kA/m^2$ will be evaluated.

The possibilities offered by ceramic diaphragms will also be evaluated, and an attempt will be made to improve them.

3. Status

Just started.

COMMISSION OF THE EUROPEAN COMMUNITIES ENERGY R & D PROGRAMME Objective: Production and utilization of hydrogen	Project or Sector: Electrolytic production of hydrogen

Title:

Development of new electrocatalytic substances for advanced electrolysis.

Duration: 12 months Period: 1/1/77 - 31/12/77	Contract No: 069-76 EHF Project No: H/B18(F)

Contractor: 1) I.F.P. - Institut Français du Pétrole
2) S.R.T.I. - Sté. de Recherches Techniques et Industrielles

Address: 1) 4 Avenue du Bois Préau, F-92500 Rueil Malmaison
2) Route de Guyancourt, F-78530 BUC

Head of Project: 1) Mr. B. Sale   2) Mr. Tribout

Description of research work

I. Objectives (aims)

To find catalysts capable of improving the water electrolysis in KOH (at about 30%) at temperatures around 150°C and moderate pressure.

2. Work Programme

Cathodic materials of the $Ni(S)$, $Ni(Zn)$ series will be improved. Their life and poisoning behaviour will be evaluated.

Mixed oxides will be studied for the anodic catalysis, in particular, the p-type semiconducting oxides, having low bonding energy for the surface oxygen. Only non-porous structures will be studied.

Detailed examination of the catalysts will be done: conductivity, state of surface, electrochemical behaviour, and chemical stability at 150°C. The satisfactory products will be tested at 20 bars up to 160°C, and at current densities up to $15\,kA/m^2$. Life testing will be done up to 4,000 hours, poisoning will be tested also. The catalyst will always be characterized after the tests.

3 Status

A good many samples of catalytic alloys have been prepared and are being tested in laboratory cells.

Total cost: 　FF.　1,315,827 　u.a.　236,915	E.C. Contribution:　50% 　FF.　657,914 　u.a.　118,458

COMMISSION OF THE EUROPEAN COMMUNITIES ENERGY R & D PROGRAMME Objective: Production and utilization of hydrogen	Project or Sector: Electrolytic production of hydrogen

Title:

Synthetic diaphragms and membranes for the electrolytic
production of hydrogen.

Duration: 12 months Period: 4/4/77 - 3/4/78	Contract No: 148-76 EHI Project No: H/B29(I)

Contractor: Politecnico di Milano, Istituto di Chimica Industriale

Address: Piazza Leonardo da Vinci, 32, I-20133 Milano

Head of Project: Prof. L. Giuffrè

Description of research work

I. Objectives (aims)

To synthesize specific organic copolymers to be used as diaphragms
or active membranes in electrolytic cells.

2. Work Programme

The copolymers to be obtained will contain active polar groups.   These
will be introduced either by standard techniques of lateral chain
addition on polymers or by copolymerisation of suitable blends of
monomers.

The polymers obtained with both preparation methods will be tested for
the following properties: concentration of the active groups, ionic
conductivity up to $1 \, A/cm^2$, ionic and dimensional stability during
electrolysis in KOH 30% at $90^{\circ}C$, stability at up to $130^{\circ}C$.

The best products will be completely evaluated, specially in electrolytic
cells at $80^{\circ}C$ for 1,000 hours and at current densities between 0.1 and
$1 \, A/cm^2$.   The potential interest and probable industrial cost of the
best polymer produced will be evaluated when the work is finished.

3. Status

Just started.

Total cost:
    Lit.   60,207,000
    u.a.   96,331
	E. C. Contribution: 68%
	    Lit. 40,932,000
	    u.a. 65,491

COMMISSION OF THE EUROPEAN COMMUNITIES ENERGY R & D PROGRAMME **Objective:** Production and utilization of hydrogen	**Project or Sector:** Electrolytic production of hydrogen

Title:

Optimization of gas evolving teflon bonded porous electrodes.

Duration: 12 months **Period:** 1/11/76 - 31/10/77	**Contract No:** 161-76 EHUK **Project No:** H/E30(UK)

Contractor: The City University (Department of Chemistry)

Address: St. John Street, London EC 1V 4PB

Head of Project: Prof. A.C.C. Tseung

Description of research work

I. Objectives (aims)

To study the teflon bonded anodic catalysts and optimize their preparation. To study the catalysts for the cathode.

2. Work Programme

For the anode the catalyst will be $NiCo_2O_4$. Several preparation methods giving various grain size and aggregation will be investigated as well as the relative quantity of teflon for bonding. Physical qualities will be measured (crystal structure, pore volume, BET surface area etc.).

Tests in electrolytic cells will be done in 30-60% KOH solutions at $70-110^{\circ}C$ and 1-10 bars. The best results will be checked in long runs (up to 6 months) to evaluate also the catalyst stability.

For the cathode: nickel black, nickel sulphide, Co molybdate and similar catalysts will be checked against the usual nickel screen. Resistance of the catalysts to oxidation during shut-down will be checked. Recommendations and prospective developments will be submitted at the end of the work.

3. Status

Just started

Total cost: £ 22,071 u.a. 52,971	**E.C. Contribution:** 79.26% £ 17,494 u.a. 41,985

COMMISSION OF THE EUROPEAN COMMUNITIES ENERGY R & D PROGRAMME **Objective:** Production and utilization of hydrogen	**Project or Sector:** Electrolytic production of hydrogen

Title:
Gas bubble formation during the electrolysis of water under forced convection conditions.

Duration: 12 months **Period:** 24/5/77	**Contract No:** 182-77 EHNL **Project No:** H/B24(NL)

Contractor: Technische Hogeschool te Eindhoven, Laboratorium voor Elektrochemie, Afdeling der Scheikundige Technologie

Address: Postbus 513, Eindhoven, The Netherlands

Head of Project: Prof. E. Barendrecht

Description of research work

I. Objectives (aims)

To study gas bubble formation during the electrolysis of water.

2. Work Programme

The formation of gas bubbles will be studied as a function of the geometry and relative arrangements of electrodes and diaphragms and for various types of electrodes.

The growth and evolution of the bubbles will be studied as a function of the electrolyte flow-rate, bubble loading, electrode structure and surface roughness, catalytic activity.

The effect of the bubbles on the voltage drop across the cell will be studied and empirical relations derived for practical cell configurations.

3. Status

Just started.

Total cost: Fl. 279,477 u.a. 77,204	**E.C. Contribution:** 50% Fl. 139,739 u.a. 38,602

Project C

Utilization of hydrogen

COMMISSION OF THE EUROPEAN COMMUNITIES ENERGY R & D PROGRAMME Objective: Production and utilization of hydrogen	Project or Sector: Utilization of hydrogen

Title:

Hydrogen storage in metallic hydrides.

Duration: 12 months Period: 15/9/76 - 14/9/77	Contract No: 041-76 EHF Project No: H/C15(F)

Contractor: Commissariat à l'Energie Atomique, Centre d'Etudes Nucléaires de Grenoble

Address: BP 85, F-38041 Grenoble Cédex

Head of Project: Mr. P. Perroud

Description of research work

I. Objectives (aims)

To improve metal hydrides and increase their stability and reaction kinetics.

2. Work Programme

The work will be concentrated mainly on magnesium hydride.   The influence of alloying elements on the storage capacity, kinetics and aging will be studied.

The effect of impurities in the feed gas and of cycling on the performances and the physical stability of the material will be examined.

The most interesting alloys will be tested in a small reservoir to gather the necessary data for the design of bigger reservoirs (specially for heat transfer to and from the hydride).

3. Status

Equipment is being set up.

Total cost:
    FF.   418,200
    u.a.   75,297
	E.C. Contribution: 50%
	    FF. 209,100
	    u.a. 37,649

COMMISSION OF THE EUROPEAN COMMUNITIES ENERGY R & D PROGRAMME Objective: Production and utilization of hydrogen	Project or Sector: Utilization of hydrogen

Title:

    Study of special thermal and chemical uses of hydrogen.

Duration: 12 months Period: 4/11/76 - 3/11/77	Contract No: 043-76 EHF Project No: H/C28(F)

Contractor: Gaz de France, Direction des Etudes et Techniques Nouvelles

Address: 23, Rue Philibert Delorme, F-75017 Paris

Head of Project: Mr. J. Pottier

Description of research work

I. Objectives (aims)

To define the industrial branches in which the use of hydrogen is economically feasible.

2. Work Programme

The technico-economic possibilities of hydrogen utilization will be examined and discussed with the interested industrial parties (not including ammonia synthesis and petrochemistry).

All the necessary information will be collected; the potential development of hydrogen uses in each branch will be studied.

A global evaluation of the potential hydrogen market for each promising industrial application in France will be carried out.

3. Status

Data collection has commenced.

Total cost:
  FF. 441,000
  u.a. 79,402
	E.C. Contribution:
	  FF. 220,500
	  u.a. 39,701

COMMISSION OF THE EUROPEAN COMMUNITIES ENERGY R & D PROGRAMME Objective: Production and utilization of hydrogen	Project or Sector: Utilization of hydrogen

Title:

Safety techniques for future hydrogen technology in Europe.

Duration: 12 months	Contract No: 046-76 EHD
Period: 21/9/76 - 20/9/77	Project No: H/C22(D)

Contractor: Dornier System GmbH.

Address: Postfach 1360, D-7990 Friedrichshafen

Head of Project: Dr. E. Wenk

Description of research work

I. Objectives (aims)

To collect and analyse all data and information on the safety aspects of hydrogen use.

2. Work Programme

Data on the hazards connected with the use of hydrogen and possible safeguards will be collected from literature. Data from industrial experience with hydrogen and the corresponding safety measures in use will be collected.

The safety regulations in European countries will be analysed. The risks involved in plants, transport, storage and consumption of hydrogen will be evaluated.

3. Status

Literature surveys and inquiries have been initiated.

Total cost:	E.C. Contribution: 100%
DM. 204,689	DM. 204,689
u.a. 55,926	u.a. 55,926

COMMISSION OF THE EUROPEAN COMMUNITIES ENERGY R & D PROGRAMME Objective: Production and utilization of hydrogen	Project or Sector: Utilization of hydrogen

Title:

Hydrogen storage with cryoabsorbers.

Duration: 12 months Period: 13/9/76 - 12/9/77	Contract No: 057-76 EHD Project No: H/C34(D)

Contractor: Deutsche Forschungs- und Versuchsanstalt für Luft- und Raumfahrt e.V. (DFVLR)

Address: Linder Höhe, Postfach 90 60 58
D-5000 Köln 90

Head of Project: Prof. W. Peschka

Description of research work

I. Objectives (aims)

To evaluate the materials of potential use for cryogenic hydrogen storage.

2. Work Programme

Materials already known to be cryogenic hydrogen absorbers will be evaluated. Their absorption capacity, kinetics and cost will be studied.

This will include experimental measurements which should also examine the life time of the absorbers and the influence of water or other impurities in hydrogen.

The preliminary design of a storage unit will be established. This will include the design of the containers of cryogenic machines, the cold heat exchangers and the storage of negative calories.

The performances of this plant will be evaluated. The cost of storage will be compared with other types of storage for hydrogen (compressed, liquified, combined in hydrides).

3. Status

An initial cost evaluation has been done. Some experimental measurements have been carried out.

Total cost: DM. 366,200 u.a. 100,055	E. C. Contribution: 50% DM. 183,100 u.a. 50,027

COMMISSION OF THE
EUROPEAN COMMUNITIES
 	
ENERGY R & D PROGRAMME
 	
Objective: Production and
      utilization of hydrogen
	Project or Sector:
	
	    Utilization of hydrogen

Title:

    Optimization studies of steels to be eventually exposed in
    hydrogen environments.

Duration: 12 months Period: 11/10/76 - 10/10/77	Contract No: 064-76 EHF Project No: H/C42(F)

Contractor: Sté. Creusot-Loire S.A., Usine du Creusot

Address: 60, Rue Clemenceau, F-71208 Le Creusot

Head of Project: Mr. J. Dollet

Description of research work

  I. Objectives (aims)

  To evaluate the behaviour of steels in hydrogen and optimize their
  heat treatments for the best resistance to hydrogen embrittlement.

  2. Work Programme

  Three different families of steel will be examined.   The influence
  of heat treatment on their sensitivity to hydrogen will be investigated.
  The best treatment will be adopted for evaluation in greater detail the
  influence of the hydrogen purity and of the temperature on the
  embrittlement.

  The rate of diffusion of hydrogen through the metal will be measured and
  deferred rupture tests will be carried out.   The influence of factors
  such as gas pressure and metal surface quality will be investigated.

  3. Status

  Procurement of steels and definition of the heat treatments has been
  done.

Total cost:
 	
    FF.   288,300
    u.a.   51,909
	E.C. Contribution: 50%
	
	    FF. 144,150
	    u.a. 25,954

COMMISSION OF THE EUROPEAN COMMUNITIES ENERGY R & D PROGRAMME **Objective:** Production and utilization of hydrogen	**Project or Sector:** Utilization of hydrogen

Title:

Technico—economic study of the use of hydrogen as a reducing agent in steelmaking.

Duration: 18 months **Period:** 1/1/77 - 30/6/78	**Contract No:** 065-76 EHF **Project No:** H/C53(F)

Contractor: Université de Metz, U.E.R. Sciences Juridiques, Economiques et Sociales, Départ. de Gestion et d'Economie Industrielle

Address: Ile du Saulcy, F-57000 Metz

Head of Project: Mr. Y. Gousty

Description of research work

I. Objectives (aims)

To evaluate the possibilities of using hydrogen for the reduction of iron ores.

2. Work Programme

A synthetic review of the known reduction processes will be done and their possible developments will be discussed.

The possibilities of using hydrogen will be analysed from the specific technico—economic point of view of European industry. The critical price at which hydrogen will be economically interesting will be evaluated.

Various processes will be considered and the influence of the important economic parameters will be taken into account.

3. Status

Just started.

Total cost: FF. 99,850 u.a. 17,978	**E.C. Contribution:** 50 % FF. 49.926 u.a. 8.988

COMMISSION OF THE EUROPEAN COMMUNITIES ENERGY R & D PROGRAMME Objective: Production and utilization of hydrogen	Project or Sector: Utilization of hydrogen

| Title:

The use of non-fossil hydrogen in coal conversion. ||

Duration: 12 months Period: 7/10/76 - 6/10/77	Contract No: 068-76 EHUK Project No: H/C57(UK)

Contractor: National Coal Board

Address: Stoke Orchard, Cheltenham, Glos. GL 52 4RZ, U.K.

Head of Project: Dr. J. Gibson (Coal Research Establishment, Stoke Orchard)

Description of research work

I. Objectives (aims)

To review the economic feasibility of using hydrogen in coal conversion processes.

2. Work Programme

Three processes for coal conversion will be considered. The analysis will give estimates of the amount of hydrogen needed, the process developments required, the thermal efficiency, the yields and the carbon utilization obtained.

The economics of the processes will also be evaluated, giving the maximum hydrogen price at which these processes will be economically feasible.

3. Status

The analysis of the three selected processes (SNG production, methanol production, coal liquefaction) has been started.

Total cost: £ 31,420 u.a. 75,408	E.C. Contribution: 50% £ 15,710 u.a. 37,704

COMMISSION OF THE EUROPEAN COMMUNITIES ENERGY R & D PROGRAMME **Objective:** Production and utilization of hydrogen	**Project or Sector:** Utilization of hydrogen

Title:

Comparison of the relative merits of hydrogen, methanol and synthetic methanol-based fuels for automotive propulsion.

Duration: 12 months **Period:** 1/1/77 - 31/12/77	**Contract No:** 070-76 EHF **Project No:** H/C30 + 31(F)

Contractor: 1) Commissariat à l'Energie Atomique (CEA-CENSACLAY)
2) Institut Français du Pétrole (I.F.P.)

Address: 1) 29-33 rue de la Fédération, Paris 7e
2) 1-4 avenue du Bois Préau - 92502 Rueil Malmaison

Head of Project: 1) Mr. Gelin; 2) Mr. Leprince

Description of research work

I. Objectives (aims)

To analyse the cost structure associated with the distribution and storage of hydrogen and methanol for the propulsion of motor-cars.

2. Work Programme

Methanol: examination of the modifications which the present petrol distribution system would require to be suitable for methanol; evaluation of the cost of the distribution of methanol;

Hydrogen: analysis of the technical solutions required for the storage, transmission and distribution of hydrogen, including service-station equipment. The safety requirements will be taken into account in the evaluation.

The investments will be estimated.

The cost and energy balance of hydrogen distribution will be calculated.

3. Status

Just started.

Total cost: FF. 508,000 u.a. 91,466	**E.C. Contribution:** 50% FF. 254,000 u.a. 45,733

COMMISSION OF THE EUROPEAN COMMUNITIES **ENERGY R & D PROGRAMME** **Objective:** Production and utilization of hydrogen	**Project or Sector:** Utilization of hydrogen

Title:

Study of the possibility of the conversion to hydrogen use of the natural-gas pipeline network.

Duration: 12 months **Period:** 1/1/77 - 31/12/77	**Contract No:** 160-76 EHF **Project No:** H/C64(F)

Contractor: Gaz de France, Direction des Etudes et Techniques Nouvelles

Address: 23, Rue Philibert Delorme, F-75017 Paris

Head of Project: Mr. S. Bezin

Description of research work

I. Objectives (aims)

To study the possibility of transporting hydrogen in existing natural-gas pipelines.

2. Work Programme

The necessary information will be collected, especially on the possible weak points due to use or to other causes. The following data will be gathered in detail: type, origin and treatments of tubes used; operative procedures (type of welding, cold work etc.); type of gas transported; stress during operation; actual conditions of the internal surfaces, corrosion, defects etc.

The tubes dismantled for various reasons will be examined in detail. Experimental tests will be done at the same time for testing the delayed rupture under the influence of hydrogen. The best type of test and the most appropriate testing conditions will be evaluated during these tests.

The materials examined will be the steel types: X42, X52 and X65, either as delivered or in welded conditions.

3. Status

Just started.

Total cost: FF. 428,500 u.a. 77,152	**E.C. Contribution:** 50% FF. 214 250 u.a. 38,576

Solar energy

Even in European latitudes, solar energy can contribute towards the
saving of a considerable amount of energy stemming from exhaustible
sources.   In the domestic sector (heating and cooling) there is even
some prospect of savings in the short term, whereas other methods
(photovoltaic and thermodynamic conversion, biomass transformation,
"artificial" photochemistry etc.) should be considered as long term
options.

Many of the basic principles and techniques to be used have already
been demonstrated on a research scale.   Further R & D is necessary
in order to develop technologies which could economically compete with
conventional energy sources.

As can be seen below and from the description of the individual
contracts, the EC programme copes with all major conversion methods
of solar radiation into conventional forms of energy.   Quite some
emphasis is put on those methods particularly suitable for the
relatively unfavourable climatic conditions of the Community countries.
For this reason, the largest single fraction of the programme is
devoted to photovoltaic conversion.   Production of organic matter as
well as production of hydrogen by photochemical or photobiological
processes also belong to this category.   Due attention is given to
heat collectors and their application for water heating, house heating
and cooling in order to further stimulate the industry of member
countries emerging in this field.

The possibility of generating electricity in a central helioelectric
power plant – a promising technology for the southern regions of the
Community – is being actively investigated.

In practical terms, the work carried out or under way in the framework
of the programme can be subdivided as follows:

A.   Heat collectors and their applications to dwellings

As the problem of storage of low temperature heat is not yet solved
satisfactorily, a number of contracts have been placed in which new
ideas for storage are investigated further and in which also other
questions related to the storage problem are studied.

Under the more general heading of "systems studies" some work has been
sponsored on mathematical models, economic analysis, optimization
studies, simulation and design of systems.   In the first phase, no
contracts have been issued on collector development and on solar
cooling devices.

B. Power production (helioelectric power plants)

In 1976 the system definition study for a 1 MW (el) central receiver helioelectric power plant has been carried out jointly by four contractors. On the basis of the results of this study it is now intended to build a full scale pilot plant, after having undertaken the necessary pre-development work and test of prototype components to be tackled still in 1977.

C. Photovoltaic conversion

A very broad range of topics has been selected for work in the first phase. In the field of silicon cells effort is devoted to optimization of classical manufacturing processes as well as to new techniques. Work on Cd-S cells is, of course, prevailing in those contracts dealing with II-VI compounds, but also other materials are studied.

Further studies concern various problems encountered with Gallium-arsenide cells and concentrator problems as well as questions related to large scale production and encapsulation of cells. At the time when these contracts were selected it was not possible (and it still is not) to state which of the technologies and cell materials under discussion would best match the requirements of terrestrial applications.

D. Photochemical, photoelectrochemical and photobiological processes

The contracts placed under this project are all aimed at supporting leading laboratories in the EC countries already engaged in photosynthesis and photochemical research. The work undertaken is mostly related to the production of hydrogen (or other fuels) by photochemical dissociation of water and in particular to the study of oxidation-reduction reactions across membranes.

E. Photosynthetic production of organic matter (Biomass)

In this field, practical and theoretical work has been undertaken on different aspects related to the use of straw as an energy source.

One contract concerns selecting and growing convenient species of fast growing trees (short rotation forestry), which will be used for power production in an existing thermal plant heated actually with peat.

F. Solar radiation data

A detailed determination of the real needs of users and research labs and an inventory of work already down elsewhere has been undertaken in the first phase. No contracts have been placed before the closing date of this report. Intense activities of a specialized working group have led to the clear identification of actions to be started in 1977: editing of radiation atlases and data books, development of climatic reference years, improvement of the existing measurement network etc.

SUMMARY AND BREAKDOWN
OF FUNDING

Objective: Solar Energy

Project	Number of contracts(*)	Total cost u.a.	E. C. contribution u.a.	Number of proposals under negotiation
A	22	1,746,080	706,194	-
B	4	196,760	136,924	-
C	26	3,976,879	1,414,870	4
D	19	831,422	397,272	-
E	5	345,974	176,791	1
F	-	-	-	-
Total	76	7,097,115	2,832,051	5

(*)Signed both by the Commission and the Contractor or sent for
signature to the Contractor

Project A

Solar heat collectors and their
application to dwellings

COMMISSION OF THE EUROPEAN COMMUNITIES ENERGY R & D PROGRAMME Objective: Solar Energy	Project or Sector: Solar heat collectors and their application to dwellings

Title:

Realization and management of solar heating systems
for space and somestic water heating in low rise buildings.

Duration: 9 months Period: 1/10/76 - 30/6/77	Contract No: 105-76 ESI Project No:

Contractor: Fiat S.p.A. (Centro Ricerche)

Address: Strada Torino, 50, - 10043 Orbassano (TO), Italy

Head of Project: Ing. Armando Campanile

Description of research work

I. Objectives (aims)

The research programme will be carried out on some low rise apartments
being part of suburban residential houses of the same type which make
them really comparable. The aim of the research is:

- to provide a facility for on site characterization of solar
  collectors produced by different manufacturers;

- to collect year round data referred to heating systems incorporating
  different energy/conservative modifications (solar equipment improved
  comfort control) and to evaluate separately the benefits of each of
  these modifications as compared to conventional systems. Pre-
  disposition works will be done in four houses, each of them having
  six flats, in this way:

No. 2 house predisposed for solar heating, with larger radiators and
p/i/d thermoregulation systems;

No. 1 house with radiators and regulation as above, but with
conventional source of heating;

No. 1 house with radiators rated for operation of higher temperature
and heated by the conventional system with proportional regulation.

../..

Total cost: Lit. 43,700,000 u.a. 69,920	E.C. Contribution: 50% Lit. 21,850,000 34,960

COMMISSION OF THE EUROPEAN COMMUNITIES ENERGY R & D PROGRAMME Objective: Solar Energy	Project or Sector: Solar heat collectors and their application to dwellings

2. Work Programme

The research programme is based on the realization of flat-plate solar systems for heating of living quarters and domestic water in condominium complexes of six flats, paired on three stories.

Broad system specifications for every condominium:

- Volume of heat space   1800 cu.m
- Max thermal power requirement   35,000 Kcal/hr.
- Radiator inlet water temperature 60^0
- Water temperature loss across radiator 10^0C
- Solar collector total surface area   100 sq.m.
- Total heat storage tank volume   15 cu.m.
- Min. outdoor design temperature 0^0C
- Room temperature 20^0C
- Yearly average solar radiation   350 W/sq.m.

3. Status

The first phase, concerning the design, is being performed.
A first design situation including the building thermal losses has been examined, and a first working test for TRNSYS simulation programme has been performed.

COMMISSION OF THE EUROPEAN COMMUNITIES ENERGY R & D PROGRAMME Objective:　Solar Energy	Project or Sector: Solar heat collectors and their application to dwellings

Title:

Numerical simulation of the thermal behaviour of a block of flats with mixed heating of air and electricity.

Duration:　　8 months Period:　1/1/77　-　31/8/77	Contract No:　　114-76　ESF Project No:

Contractor:　　Société Bertin & Cie

Address:　　Allée Gabriel Voisin - 78370 Plaisir (Yvelines), France

Head of Project:　　R. Grossin

Description of research work

I. Objectives (aims)

To demonstrate the performance of a solar heating system for a block of flats.

2. Work Programme

2.1. The development of a simulation model;

2.2. The definition of the performance of this heating system in three different types of climate;

2.3. Economical assessment of this heating system.

3. Status

The simulation model is ready for testing.　Meteorological hourly values are available.

Total cost: 　FF.　259,143 　u.a.　46,658	E.C. Contribution:　53% 　FF.　138,599 　u.a.　24,955

COMMISSION OF THE EUROPEAN COMMUNITIES ENERGY R & D PROGRAMME Objective: Solar Energy	Project or Sector: Solar heat collectors and their application to dwellings

Title:

A study of the storage of energy using the latent heat of fusion of a substance absorbed by a support of the chromatographical type. Comparison with more conventional systems.

Duration: 24 months Period: 15/12/76 - 14/12/78	Contract No: 115-76 ESF Project No:

Contractor: Société Nationale Elf Aquitaine (S.N.E.A.)

Address: Tour Aquitaine, 92080 Paris-La Defense, France

Head of Project: Mr M. Rone

Description of research work

I. Objectives (aims)

The development of phase-change material for heat storage, whereby the p.c.-material is absorbed on a solid absorber in the form of granules.

2. Work Programme

Theoretical and experimental study concerning:

2.1. the dynamic temperature profiles in the heat store;

2.2. the static behaviour of the heat store;

2.3. a comparison with more classical heat stores;

2.4. optimization of heat exchange in particular and of the system behaviour in general.

3. Status

The investigation has started to compare a heat store consisting of an absorbent impregnated with a paraffin and a heat store consisting of a bed of encapsulated paraffin balls.

Total cost: FF. 267,695 u.a. 48,199	E.C. Contribution: 50% FF. 160,617 u.a. 28,919

COMMISSION OF THE EUROPEAN COMMUNITIES ENERGY R & D PROGRAMME Objective: Solar Energy	Project or Sector: Solar heat collectors and their application to dwellings

Title:

Storage of the solar energy and waste heat of industrial processes in chemical bonds in the form of thermodecomposable compounds.

Duration: 18 months Period:	Contract No: 116-76 ESB Project No:

Contractor: Faculté Polytechnique de Mons

Address: Rue de Houdain, 9, 7000 Mons, Belgium

Head of Project: M. J. Bougard

Description of research work

I. Objectives (aims)

Search of exothermic chemical reactions produced from reagents stored apart at room temperature. The reactions must have the following characteristics:

(a) heat delivered (Q) > 0.1 cal/m^3 of stored reagents;

(b) the reversal of the reaction may occur below $120°C$;

(c) in order to diminish the volume of the storage tank, the search will be turned to chemical reactions in which one of the reagents will be the water of the air;

(d) if it is possible, liquid systems will be chosen;

(e) cost of products;

(f) toxicity;

(g) time stability of product.

2. Work Programme

Phase 1.

– Bibliography study and thermodynamic analysis of the various chemical reactions corresponding to the scope of the storage;

– Experimental study of retained systems: calorimetric measurements; reversibility study; study of separation processes;

../..

Total cost: BF. 2,579,850 u.a. 51,597	E.C. Contribution: 50% BF. 1,289,925 u.a. 25,799

COMMISSION OF THE EUROPEAN COMMUNITIES ENERGY R & D PROGRAMME Objective: Solar Energy	Project or Sector: Solar heat collectors and their application to dwellings

2. Work Programme cont/..

- Proposition of one of several systems consistent with the constraints of chemical engineering.

Phase 2.

- Complementary thermodynamic studies for the realisation of a pilot unit;

- Project study of a complete pilot unit including a complete cycle: reaction, decomposition, separation;

- Construction of the pilot unit;

- Experiment on the pilot unit;

- Conclusion;

- Economic evaluation.

3   Status

Being signed.

COMMISSION OF THE EUROPEAN COMMUNITIES ENERGY R & D PROGRAMME Objective: Solar Energy	Project or Sector: Solar heat collectors and their application to dwellings

Title:

    Solar heating and interseasonal storage for a group of dwellings.

Duration: 12 months Period: 1/9/76 - 31/8/77	Contract No: 117-76 ESB Project No:

Contractor: Advanced Technology Research and Application Company (A.T.R.A.C.)

Address: 98, Avenue Tervuren, 1040 Bruxelles, Belgium

Head of Project: Mr. C. Heusquin

Description of research work

  I. Objectives (aims)

    1.1. An inventory of the existing solar energy storage projects and projects under investigation;

    1.2. Selection of the best suited projects for a medium term large scale application in European areas around the 50th parallel north;

    1.3. An optimization study of the installations;

    1.4. Technical economical assessment of the selected projects.

  2. Work Programme

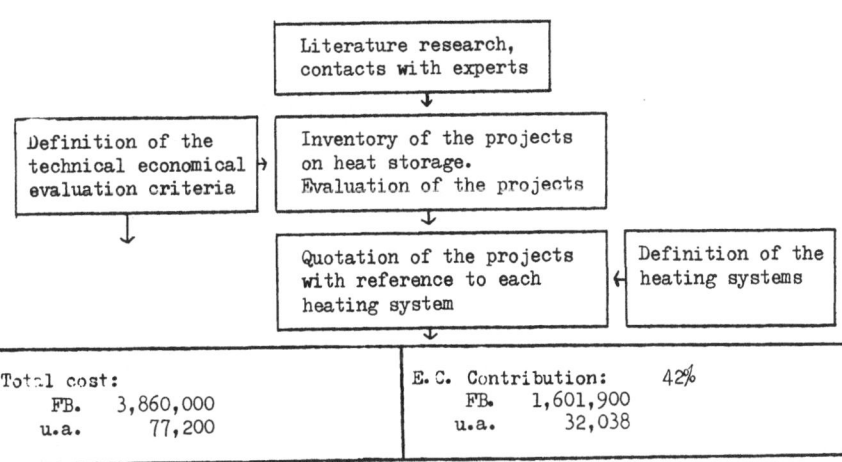

Total cost:
   FB.   3,860,000
   u.a.   77,200
	E.C. Contribution: 42%
	   FB. 1,601,900
	   u.a. 32,038

COMMISSION OF THE EUROPEAN COMMUNITIES ENERGY R & D PROGRAMME Objective: Solar Energy	Project or Sector: Solar heat collectors and their application to dwellings

2. Work Programme cont/..

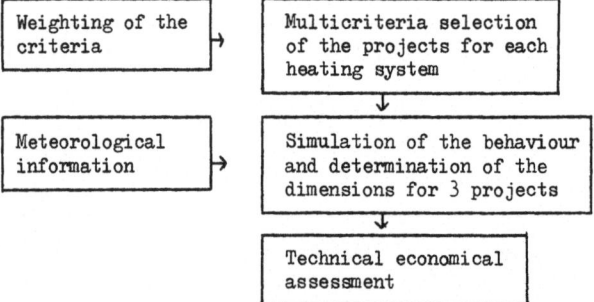

3. Status

The structures for the evaluation of the storage projects have been developed. The assessment criteria, the quotation scales as well as the weighting for each criteria have been elaborated.

COMMISSION OF THE EUROPEAN COMMUNITIES ENERGY R & D PROGRAMME Objective: Solar Energy	Project or Sector: Solar heat collectors and their application to dwellings

Title:

A study of the seasonal heat storage systems at about 80°C.

Duration: 18 months Period: 1/10/76 - 31/3/78	Contract No: 118-76 ESF Project No:

Contractor: C.E.A.

Address: B.P. No. 2 - (91190) Gif-sur-Yvette, France

Head of Project: J. Bernard - S.E.E.N. - bâtiment 24

Description of research work

I. Objectives (aims)

The behaviour of an actual energy storage unit is studied by means of a small scale experimental unit (4 m3), the walls of which are jacketed so as to simulate various thermal boundary conditions.

Different fillings (pebbles, rocks, etc.) are tested under several heating and cooling programmes. Experimental results are compared with a computer model. A technical and economic study of an energy storage unit for dwellings can then be made.

2. Work Programme

(a) Study of the experimental rig and setting;

(b) Bibliography and first model;

(c) Testing of the experimental rig (water tightness, heaters, connections, etc..);

(d) Testing of several fillings under various heating and cooling programmes. Comparison of experimental and computed results. Choice of a kind of filling;

(e) More precise investigation of the chosen filling. Study of the behavior of the storage unit for various thermal boundary conditions. Evaluation of heat losses. Derivation of a more complete model;

(f) Use of a storage unit. Determination of a suitable network of water inlets/outlets.

Total cost:
   FF.   888,600
   u.a.   159,993
	E.C. Contribution: 26%
	   FF. 277,243
	   u.a. 49,918

COMMISSION OF THE EUROPEAN COMMUNITIES ENERGY R & D PROGRAMME Objective: Solar Energy	Project or Sector: Solar heat collectors and their application to dwellings

3. Status

At end March 1977.   Points a, b, and c are executed.   The first testing of fillings will be made when we get the 100 channels data logger (last week of March).

COMMISSION OF THE EUROPEAN COMMUNITIES ENERGY R & D PROGRAMME Objective: Solar Energy	Project or Sector: Solar heat collectors and their application to dwellings

| Title:

    Construction of a complete all season conditioning with solar energy of an office building of 200 mc and its operation. ||

Duration: 30 months Period: 10/1977 - 03/1979	Contract No: 121-76 ESI Project No:

| Contractor: Nuovo Pignone S.p.A.

Address: Firenze - 2, Via F. Matteucci, Italy

Head of Project: Ing. Gian Pietre Ferrari Aggradi ||

Description of research work

  I. Objectives (aims)

Study of a type of building allowing ease of integration with a solar plant. Conditioning, also in summer, using an absorbing plant check for economy of availability of a perpetual non-polluting source of energy as the sun is; collection of experimental data on isolation, weather conditions, performance of the single component and the whole system, operating and maintenance expenses; data processing and optimisation of the solar-traditional plant.

  2. Work Programme

October '76 - December '76	:	Building and plant design
January '77 - February '77	:	Orders and materials supply
March '77 - June '77	:	Building construction plant assembly and testing
July '77 - March '79	:	Experimentation
		Data collection
		Data analysis
		Optimisation studies for the solar system.

  3. Status

  (i) Studies   : Building and plant design (for all its working conditions) are accomplished; ../..

Total cost:
    Lit. 100,108,000
    u.a. 160,173
	E.C. Contribution: 16%
	    Lit. 15,625,000
	    u.a. 25,000

COMMISSION OF THE EUROPEAN COMMUNITIES ENERGY R & D PROGRAMME Objective: Solar Energy	Project or Sector: Solar heat collectors and their application to dwellings

3. **Status** cont/..

(ii)   Construction   :   Reinforced concrete structure is over (ground work and basement);   assembly of iron structure has started, and after that the assembling of the pre-fabricated elements and the plant will take place;

(iii)   Instrumentation : Most components of the solar system, of plant regulation instruments and of automatic data acquisition system are already in place.

COMMISSION OF THE EUROPEAN COMMUNITIES ENERGY R & D PROGRAMME Objective: Solar Energy	Project or Sector: Solar heat collectors and their application to dwellings

Title:

Latent heat of fusion storage for solar systems.

Duration: 15 months Period: 1/9/76 - 30/11/77	Contract No: 129-76 ESD Project No:

Contractor: Brown, Boveri & Cie Aktiengesellschaft

Address: 6800 Mannheim 1, Germany

Head of Project: H. Birnbreier

Description of research work

I. Objectives (aims)

The main objective is the investigation of latent heat of fusion energy storage at low storage temperatures for solar systems with heat pumps.

Latent heat of fusion storage will be compared with sensitive heat storage in water. Substantial reductions in storage volume to about 25% of that for an equivalent water capacitor and lower investment costs are expected.

2. Work Programme

(a) Selection of suitable storage materials by laboratory tests (Hexadecane, operating temperature: $18^\circ C$; water/ice/operating temperature: $0^\circ C$);

The selection of these media is the result of an optimization with respect to a high coefficient of performance for the heat pump on the one hand and a high efficiency for the solar collectors on the other hand;

(b) Design and construction of a latent heat of fusion capacitor with hexadecane and water/ice. The performance data (i.e., storage capacity, energy input and energy withdrawal rate) correspond to those of an 8 m^3 water capacitor. Installation in the solar system of the BBO-Solar-House;

(c) Operation and performance of a solar heating and cooling system utilizing a latent heat of fusion storage subsystem and an equivalent water storage subsystem;   ../..

Total cost:
    DM.   266,390
    u.a.   72,784
	E.C. Contribution: 50%
	    DM. 133,195
	    u.a. 36,392

COMMISSION OF THE EUROPEAN COMMUNITIES ENERGY R & D PROGRAMME Objective: Solar Energy	Project or Sector: Solar heat collectors and their application to dwellings

2. Work Programme cont/..

(c) cont/..

Both capacitors will be operated alternatively for several test periods of 2 - 4 weeks.   The results of these long term tests will lead to cost optimized design data for heat of fusion capacitors in solar systems with heat pumps.

3. Status

Based on the characteristics of solar energy systems for space heating and cooling and the selection criteria for latent heat of fusion storage materials, paraffin hydrocarbons and water/ice are very promising phase-change materials.

Different paraffin hydrocarbons and water/ice have been investigated in a laboratory test capacitor.

The main design criteria for latent heat of fusion capacitors equivalent to an 8 m^3 water capacitor have been established.

A simplified analytical model has been developed and successfully applied on the test capacitor.

Presently the latent heat of fusion storage unit, designed for a one family residence building, is being installed in the solar heating and cooling system of the BBO-Solar-House.

COMMISSION OF THE EUROPEAN COMMUNITIES ENERGY R & D PROGRAMME Objective: Solar Energy	Project or Sector: Solar heat collectors and their application to dwellings

Title:

Application of solar energy in dwellings – a technical and economical analysis.

Duration: 11 months Period: 1/10/76 - 30/9/77	Contract No: 130-76 ESD Project No:

Contractor: Messerschmitt-Bolkow-Blohm

Address: D -8000 München 80, Postfach 80 11 69, Germany

Head of Project: Dipl.-Ing. H. Schweig

Description of research work

I. Objectives (aims)

The primary objective of the study is a technical and economical analysis for the different European climates aiming at a recommendation on which system is to be installed in what country.   This includes the definition of technological problem areas to be solved in the future. The economical conditions of solar energy as applied to dwellings, in comparison to other primary energy systems, will be analysed, ending in a forecast on which part of the present primary energy can be substituted by solar energy up to the end of this century.

2. Work Programme

Analysis of present and future designs and estimation of costs. Definition of solar systems and analysis with a complex computer programme, including hot water preparation, house heating and cooling. Variation of heat transport media (water, water plus antifreeze, synthetic media, air), variation of collector types, variation of heat storage size.

Analysis of control units for proper operation.

Estimation of system cost based on the experience with three test houses and a first production series of more than 300 hot water preparation systems.

Economic comparison of solar systems with conventional units.

Definition of a technology programme for Europe.   ../..

Total cost: DM. 279,920 u.a. 76,481	E.C. Contribution: 64% DM. 179,149 u.a. 48,948

COMMISSION OF THE
EUROPEAN COMMUNITIES

ENERGY R & D PROGRAMME

Objective: Solar Energy

Project or Sector:

Solar heat collectors and their
application to dwellings

2. Work Programme cont/..

Estimation of primary energy fraction which can be substituted
by solar energy until 2000.

3. Status

A detailed simulation for Germany has been completed.   The inputs
for other countries (mainly climate data) are prepared.   The
simulation will be completed at the end of June 1977.

The cost estimation for various systems is nearly completed.   An
economic evaluation and comparison will be performed in July,
following the simulation results.   A report containing all material
will be prepared in August and September 1977.

COMMISSION OF THE EUROPEAN COMMUNITIES ENERGY R & D PROGRAMME Objective: Solar Energy	Project or Sector: Solar heat collectors and their application to dwellings

Title:

The use of solar components for space heating and water heating in dwellings: mathematical models and the development of practical guidelines.

Duration: 18 months Period: 1/7/76 - 31/12/77	Contract No: 131-76 ESD Project No:

Contractor: Fraunhofer Gesellschaft zur Förderung der angewandten Forschung e.V.

Address: Institut für Systemtechnik und Innovationsforschung, Breslauer Str. 48, 75 Karlsruhe-Waldstradt

Head of Project: Dr. Richard Denton (Institut für Systemtechnik)

Description of research work

I. Objectives (aims)

Ultimate Goal: To provide a set of information manuals for the design and proper dimensioning of solar systems using flat plate collectors for a range of prototype dwellings.

Target Groups: Architects, builders, plumbing and heating firms, manufacturers, homeowners and potential homeowners.

Background Motivation for Project: General lack of well-founded technical information and realistic cost comparisons between solar and conventional systems could lead to false expectations by the public at large; failure to meet these expectations could result in a setback in the eventual market penetration.

2. Work Programme

(a) Basic technical information:- from scientific and sales literature, from discussions with manufacturers and other workers in the field, and questionnaires.

     (i)   solar components:

         - performance and cost data at present;

         - expected price trends, possible economies of scale in manufacturing, etc.

     (ii)   meteorological data:

                                     ../..

Total cost:
     DM.   191,816
     u.a.   52,409
	E.C. Contribution: 50%
	     DM. 95,900
	     u.a. 26,202

COMMISSION OF THE EUROPEAN COMMUNITIES ENERGY R & D PROGRAMME Objective: Solar Energy	Project or Sector: Solar heat collectors and their application to dwellings

2. Work Programme cont/..

- insolation: direct and diffuse, hourly values;
- use of a stochastic model which generates the data from global insolation values on a horizontal surface;
- hourly ambient temperatures;
- wind.

(iii) home heating and hot water demands:

- heating demands based on degree-hour model;
- later work to include comparison with more realistic demand models, which incorporate time lags via z-transforms;
- exogeneous hourly hot water demand profiles.

(b) Solar system configurations:

(i) applications;

(ii) analysis of main configurations from an energy consumption standpoint;

(iii) total system costs.

(c) Static considerations:

(i) initial solar system optimization on a:

- technical basis;
- cost basis.

(d) Simulation model:

(i) system equations;

(ii) switching strategies;

- primary circuit;
- secondary circuit - depends on necessary heating medium temperatures as calculated from demand model;

(iii) computer runs using hourly data over a full year;

(iv) description of resutls in parametric form - for example, using a generalization of the "f-chart" method.

3. Status

Collection and analysis of the "basic technical information" is complete. Reports on the main aspects are currently being written.

Roughly six categories of system configurations have been chosen for detailed analysis. Simulation modelling of these configurations is in progress.

COMMISSION OF THE EUROPEAN COMMUNITIES ENERGY R & D PROGRAMME Objective: Solar Energy	Project or Sector: Solar heat collectors and their application to dwellings

Title:
 	
    Development of a thermal storage system based on
    encapsulated p.c.m.-materials.	

Duration: 12 months Period: 1/1/77 - 31/12/77	Contract No: 133-76 ESN Project No:

Contractor: TNO - TH - Institute of Applied Physics Address: Stieltjesweg 1, Delft, The Netherlands Head of Project: Ir. C. den Ouden	

Description of research work

I. Objectives (aims)

Development of a short term storage system, based on a phase change material encapsulated in a polymer, capable of storing a few days surplus of solar heat (< 200 kWh ≈ 700 MJ) at a temperature level of $35-60^\circ$C.

2. Work Programme

(a) Selecting of suitable p.c.m.-materials;

(b) Testing the thermal behaviour of a few promising encapsulated materials;

(c) Establishing the relevant parameters of the thermal storage system;

(d) Design and production of a first prototype - Preliminary tests.

3. Status

Three promising p.c.m.-materials are encapsulated that show no supercooling and promising thermal properties.   A mathematical model has been developed to predict in more detail the heat transfer mechanism inside the tss.
Input parameters for this model:
- the thermal properties of the encapsulated p.c.m.-material;
- the dimensions of the tss;
- in- and output conditions as a function of time.

Total cost:
    Fl.   287,000
    u.a.   79,282
	E.C. Contribution: 50%
	    Fl. 126,280
	    u.a. 34,884

COMMISSION OF THE EUROPEAN COMMUNITIES ENERGY R & D PROGRAMME Objective: Solar Energy	Project or Sector: Solar heat collectors and their application to dwellings

Title:

A technical and economic study of the potential use of solar energy for the provision of space heating and hot water for dwellings, using stochastic modelling.

Duration: 12 months Period: 1/10/76 - 30/9/77	Contract No: 134-76 ESEIR Project No:

Contractor: Institute for Industrial Research and Standards

Address: Ballymun Road, Dublin 9, Ireland

Head of Project: Mr. E. Kinsella

Description of research work

I. Objectives (aims)

Assessment of the potential use of solar energy for the provision of space heating and hot water for dwellings.

2. Work Programme

2.1. The development of a meteorological reference year;

2.2. Modelling of performance characteristics of system components;

2.3. Definition of energy demand patterns;

2.4. Identification of criteria related to thermal performance and cost effectiveness;

2.5. Stochastic modelling.

3. Status

Work as defined under 2.1. - 2.5. is in progress.

Total cost:
    £   41,100
    u.a. 98,640
	E. C. Contribution: 36%
	    £ 14,590
	    u.a. 35,016

COMMISSION OF THE EUROPEAN COMMUNITIES	Project or Sector:
ENERGY R & D PROGRAMME	Solar heat collectors and their application to dwellings
Objective: Solar Energy	

Title:

The solar wall - analysis, optimization, performance.

Duration: 36 months	Contract No: 135-76 ESUK
Period: 1/7/76 - 31/6/79	Project No:

Contractor: University of Leeds

Address: Leeds LS2 9JT

Head of Project: Dr. D. Fitzgerald

Description of research work

I.   Objectives (aims)

(a) Produce a mathematical model to calculate the behaviour of a given design of wall in a given climate, to agree with the observed behaviour;

(b) Build a wall to examine parameters shown to be important by the model, and to monitor the performance of the wall for at least two heating seasons;

(c) Detail the wall so as to make one version acceptable to small traditionally oriented builders, and to make another version compatible with building system.

2.   Work Programme

- Have the mathematical model working by the end of 1976;

- Build Mk I wall in the spring of 1977 and monitor;

- Using the experience of the Mk I wall, build Mk II wall(s), later in 1977;

- Monitor wall(s) in the heating seasons 1977/8 and 1978/9, not forgetting the behaviour of the wall(s) in the summer;

- Detail wall(s) for commercial application in 1978;

- Report by mid-1979.   ../..

Total cost:	E.C. Contribution: 53%
£ 22,005	£ 11,553
u.a. 52,812	u.a. 27,727

COMMISSION OF THE EUROPEAN COMMUNITIES ENERGY R & D PROGRAMME Objective: Solar Energy	Project or Sector: Solar heat collectors and their application to dwellings

3. Status

Computer model working February 1977.

Mk I wall being detailed March 1977.

COMMISSION OF THE EUROPEAN COMMUNITIES ENERGY R & D PROGRAMME Objective: Solar Energy	Project or Sector: Solar heat collectors and their application to dwellings

Title:

      Optimum design of solar energy systems by multivariat analysis.

Duration: 36 months Period: 25/5/77 — 25/5/1980	Contract No: 136-76 ESUK Project No:

Contractor: University College Cardiff

Address: P.O. Box 78, Cardiff CF1 1XP, Wales, United Kingdom

Head of Project: Prof. B. J. Brinkworth

Description of research work

I. Objectives (aims)

Development of optimisation programmes, based on a group of performance programmes, in order to arrive at a methodology for the optimimum design of solar energy systems.

2. Work Programme

1. Specification of basic quantities; e.g. those related to the scale of the system and operational restraints;

2. Preparation of the optimisation procedure;

3. Test of the optimisation programme by applying the programme to the design of a trial water heater system which will be constructed and put on test in a solar stimulator;

4. Field measurements of meteorological parameters and user-demand patterns will be extended as part of the work proposed;

5. Consideration will be given to the application of the methodology to the design optimisation of other solar energy systems.

3. Status

Just started.

Total cost:
    £   23,007
    u.a.   55,217
	E.C. Contribution: 37%
	    £ 8,512
	    u.a. 20,429

COMMISSION OF THE EUROPEAN COMMUNITIES ENERGY R & D PROGRAMME Objective:　　Solar Energy	Project or Sector: Solar heat collectors and their application to dwellings

Title:

　　　Feasibility study on a solar house heating system with a low quality thermal flow.

Duration:　　　12 months Period:　1/9/76　-　31/8/77	Contract No:　　　137-76　ESDK Project No:

Contractor:　　A/S International Solar Power Co. Ltd.

Address:　　　22 B Rosenkaeret, DK 2860, Soborg, Denmark

Head of Project:　　Mr. A. Eggers-Lura

Description of research work

I. Objectives (aims)

Design and development of a prefabricated concrete building system which incorporates a solar heating system with a low quality thermal flow.

2. Work Programme

Phase 1:

(a) Description of concept philosophy;

(b) Technical description and evaluation of existing houses, which utilize high density walls as solar collectors;

(c) Technical description of prefabricated concrete building systems, which lend themselves to adaptation with low quality solar heating systems;

(d) Technical description of solar heating system components, which lend themselves to adaptation with prefabricated concrete building systems.

Phase 2:

On the basis of the material collected, and conclusions drawn, during phase 1 of the study, investigation and analysis of the parameters and data for a prefabricated concrete building system, which incorporates a solar heating system with a low quality thermal flow.　Technical and economical evaluation of the concept.

　　　　　　　　　　　　　　../..

Total cost: 　Dk.　398,000 　u.a.　53,067	E.C. Contribution:　46.5% 　Dk.　185,070 　u.a.　24,676

COMMISSION OF THE EUROPEAN COMMUNITIES ENERGY R & D PROGRAMME Objective: Solar Energy	Project or Sector: Solar heat collectors and their application to dwellings

3. Status

The work on the project was started in May 1976.   Data for phase 1 of the study has been collected.

Construction of a solar/heatpump house started in March 1977 in Skive, Denmark.   From this house data will be obtained on various solar/heatpump components, which are included in phase 2 of the study.

Phase 2 of the study has started.

COMMISSION OF THE EUROPEAN COMMUNITIES ENERGY R & D PROGRAMME Objective: Solar Energy	Project or Sector: Solar heat collectors and their application to dwellings

Title:

Development of solar energy houses based on an integrated collector storage system combined with warm air heating.

Duration: 18 months Period: 1/10/77 - 31/3/79	Contract No: 138-76 ESN Project No:

Contractor: Stichtingen Bouweentrum en Ratiobouw

Address: Weena 700, Rotterdam Holland

Head of Project: H. van Bremen/J.M. van Heel

Description of research work

I. Objectives (aims)

The formulation of directives for the design of solar energy installations especially for low-cost housing (one-family houses and multi-storage buildings). The installations will be both for heating and hot water supply. They are based on an integrated collector/storage system combined with air-heating.

This kind of solar energy installation has already been built and a prototype was tested during the winter of 75-76.

The investigation studies the connection between thermal insulation, collector surface and storage system, all for dwellings of the usual design. The prototype installation is characterised by low investment and maintenance costs and a long life span.

2. Work Programme

(a) Analysis of the eligible systems and dwelling designs (Bouwcentrum);

(b) Optimalization of the architectural design, the insulation and the solar energy system on the basis of a calculation model (Bouwcentrum + Technisch Fysische Dienst TNO-TH);

(c) Integration of the architectural design and the system (Bouwcentrum);

../..

Total cost: Fl. 200,000 u.a. 55,249	E.C. Contribution: 45% Fl. 90,000 u.a. 24,862

COMMISSION OF THE EUROPEAN COMMUNITIES ENERGY R & D PROGRAMME Objective: Solar Energy	Project or Sector: Solar heat collectors and their application to dwellings

2. Work Programme cont/..

(d) Measurements carried out on test collectors, for heating and hot water supply (test collectors are made available by the manufacturers concerned) (Technisch Fysische Dienst TNO-TH);

(e) Cost-benefit analysis (Bouwcentrum + Technisch Fysische Dienst TNO-TH);

(f) Directives for the design of solar dwellings, on the basis of the systems concerned (Bouwcentrum + Technisch Fysische Dienst TNO-TH);

(g) Writing of the report (Bouwcentrum).

3. Status

Section (a) has been completed.

Items (b) and (c) are 50% completed; a computer model for an integrated collector/storage system has been carried out.

Arrangements for making test-collectors have been made.   Early September 1977 the first series of measurements will be realised.

A test-model of an air-heated boiler for hot water supply was made and tested during several weeks.

These investigations have not yet been finished.

COMMISSION OF THE
EUROPEAN COMMUNITIES

ENERGY R & D PROGRAMME

Objective: Solar Energy

Project or Sector:

Solar heat collectors and their
application to dwellings

Title:

Multiple solar house project - development of solar
heating for use in mass-market housing.

Duration: 9 months

Period: 5/12/76 - 14/6/77

Contract No: 139-76 ESUK

Project No:

Contractor: Polytechnic of Central London, in association with
Milton Keynes Development Corp. and John Laing Research
and Development Ltd.

Built Environment Research Group, 35 Marylebone Road,
London NW1 5LS
Head of Project: Professor R. Maw

Description of research work

I. Objectives (aims)

The overall aim of the research programme is to explore some
alternative possibilities for utilising low cost solar energy systems
for heating in the mass housing sector. A group of houses
incorporating a shared solar heating system will be built as part of
a typical housing development to investigate the practical and economic
implications.

2. Work Programme

The complete programme will consist of a number of sequential phases
comprising technical design, working drawings and construction of a
group of test houses, incorporating a shared solar heating system
followed by monitoring and analysis of the overall performance.
The programme is envisaged to last at least three years.

This first stage is a feasibility study comprising:

(a) An initial appraisal consisting of an overall survey of
alternative possible solutions to determine the most important
areas for detailed study. A series of sketch design of shared
solar heating systems and associated house structures will be
outlined to ensure realistic comparative evaluation;

(b) A technical study to assess the effect of different design and
sizes of components upon the performance, capital and running
costs of the shared solar heating systems and metering arrangements;

../..

Total cost:
£   58,475
u.a. 140,340

E.C. Contribution: 22%
£   12,864
u.a. 30,874

COMMISSION OF THE EUROPEAN COMMUNITIES ENERGY R & D PROGRAMME Objective: Solar Energy	Project or Sector: Solar heat collectors and their application to dwellings

2. Work Programme cont/..

(b)   cont/...
A report of findings together with an outlined design of the 'best' practical solution for the group of test houses will be drawn up and submitted;

(c)   The preparation of sketch drawings and preliminary specifications for the solar heating system to be incorporated into a group of about sixteen to twenty houses and subsequently to be monitored under normal occupation.

3. Status

This is a continuation of an on-going research and development programme.   A standard corporation house has already been built in Milton Keynes incorporating a solar space and domestic hot water heating system; and has been monitored over the past two years under normal family occupation.

A detailed analysis of the performance of the 'solar house' has been carried out to find out how the system could be modified to improve its performance and reduce its cost.

Arising from these studies, it appears that one of the most promising ways of improving the economic viability is to build groups of houses sharing a storage, control and supplementary heating system.

The data from the monitoring of the 'solar house' have been used in conjunction with a modified version of the NBS computer simulation programme to evaluate the performance of several alternative shared solar heating systems which have been designed.   Outline cost studies of these alternatives are now in progress.

COMMISSION OF THE EUROPEAN COMMUNITIES ENERGY R & D PROGRAMME Objective: Solar Energy	Project or Sector: Solar heat collectors and their application to dwellings

Title:

Heat storage in a solar heating system using salt hydrates.

Duration: 9 months Period: 1/11/76 - 31/7/77	Contract No: 140-76 ESDK Project No:

Contractor: Thermal Insulation Laboratory

Address: Technical University of Denmark, Building 118, DK-2800 Lyngby, Denmark

Head of Project: Vagn Korsgaard

Description of research work

I. Objectives (aims)

The aim is to develop an energy storage making use of, for example, sodium sulfate decahydrate as a storage medium, without using small tanks, and without getting separation problems. A new technique, consisting of adding extra water to the mixture and keeping it in soft stirring, will be investigated.

2. Work Programme

(a) A study of relevant literature for the subject is carried out in order to elucidate the status of relevant research and development work;

(b) Some of the most promising storage systems using the method of adding extra water to the salt hydrate will be selected. It will be a question of various salt hydrates and various solutions to the problems: super cooling, separation and heat transfer to and from storage;

(c) The selected systems will be examined carefully by means of calculations and experiments. The thermal performance, durability, approximate construction and working expenses of the storage systems will be determined.

(d) A report concerning the above-mentioned points will be prepared.

../..

Total cost: Dk. 450,300 u.a. 60,040	E.C. Contribution: 50% Dk. 225,150 u.a. 30,020

COMMISSION OF THE EUROPEAN COMMUNITIES ENERGY R & D PROGRAMME Objective: Solar Energy	Project or Sector: Solar heat collectors and their application to dwellings

3. Status

At end March: (a) and (b) of Work Programme are almost finished, (c) is just starting.

COMMISSION OF THE EUROPEAN COMMUNITIES ENERGY R & D PROGRAMME Objective: Solar Energy	Project or Sector: Solar heat collectors and their application to dwellings

Title:

Static and dynamic analysis of air solar systems, combined with a compression heat pump, for domestic heating. Theoretical and experimental investigation.

Duration: 15 months Period: 1/10/76 - 31/12/77	Contract No: 141-76 ESB Project No:

Contractor: Katholicke Universiteit Leuven (K.U.L.)

Address: Leuven, Belgium

Head of Project: Prof. Dr. Ir. W.L. Dutré

Description of research work

I.   Objectives (aims)

Analysis of combined solar energy heat pump air heating systems, using air collectors and a compression air-air heat pump.

The main objective of this first phase is to develop calculational procedures to evaluate the performance of such a combined system for some possible configurations, including the regulation criteria and to compare the overall performance to the performance of analogous systems without eat pump.

2.   Work Programme

Experimental and theoretical work will be developed simultaneously.

Experimental programme:

- measurement of air collector efficiency curves;
- measurement of heat pump characteristics for a wide range of working conditions.

Theoretical work:

- development of computer programmes for the performance calculation of the combined systems, using a climatological reference year;
- comparative calculations to evaluate the influence of the heat pump on the system performance.

3.   Status   Just started.

Total cost: FB. 2,977,400 u.a. 59,548	E.C. Contribution: 50% FB. 1,488,700 u.a. 29,774

COMMISSION OF THE EUROPEAN COMMUNITIES ENERGY R & D PROGRAMME Objective: Solar Energy	Project or Sector: Solar heat collectors and their application to dwellings

Title:

Parametric analysis of solar energy systems for room heating and cooling application.

Duration: 12 months Period: 1/7/76 - 30/6/77	Contract No: 149-76 ESI Project No:

Contractor: Ind. A. Zanussi S.p.A.

Address: Viale Treviso, 15 - 33170 Prodenone, Italy

Head of Project: dott. ing. Mario de Renzio

Description of research work

I. Objectives (aims)

(a) Evaluation of the effective efficiency of a working solar plant;

(b) Evaluation of the energetical coefficients of a solar plant as for instance: utilization coefficient of useful energy from collectors, form storage;

(c) Analitical study of the energetical data elaborated by a mathematical simulation model.

2. Work Programme

(i) Examination of the different types of solar collectors;

(ii) Evaluation of the influence of tilt and direction factors on the incoming solar energy computation;

(iii) Solar collector's efficiency evaluation with respect to tilt and direction;

(iv) Computer simulation of Susegana's plant

(a) evaluation of the optimum solar collector's surface area;

(b) comparison between computer simulated results and the experimental results;

(v) Experimental data evaluation for the solar plant at Porcia. ../..

Total cost: Lit. 57,470,000 u.a. 91,952	E.C. Contribution: 49% Lit. 28,125,000 u.a. 45,000

COMMISSION OF THE EUROPEAN COMMUNITIES ENERGY R & D PROGRAMME Objective: Solar Energy	Project or Sector: Solar heat collectors and their application to dwellings

3. Status

Parts (i), (ii) and (iv(a) have been completed.

We are at present setting up the installation of the data-logger and we expect it will be operating at the middle of June.

COMMISSION OF THE EUROPEAN COMMUNITIES ENERGY R & D PROGRAMME Objective: Solar Energy	Project or Sector: Solar heat collectors and their application to dwellings

Title:

Research on products for heat storage at low temperatures
between (5 - 125°C).

Duration: 16 months Period: 1/9/76 - 31/12/77	Contract No: 168-76 ESF Project No:

Contractor: Société Rhône-Poulenc Industries

Address: 22, Avenue Montaigne - 75008 Paris, France

Head of Project: M. F. Vachet, Centre de Recherches de DECINES

Description of research work

I. Objectives (aims)

The work is aimed at checking the practicality and economic value of
storing latent heat of fusion in chemicals in the temperature range and
operative mode compatible with solar heating of dwellings.

If suitable products are readily available from manufacturers or other
laboratories, they will be used.   If not, research will be undertaken
to develop one.   A rough cost objective of 2.5. u.a./MJ is considered
suitable for a storage "appliance" ready to be connected.

2. Work Programme

- Survey of available knowledge and products in the field;
- Heat transfer calculations to obtain a simple modelling of the
  expected thermal response of products to be prepared;
- Mass transfer calculations to determine the suitability of candidate
  containment materials in various conditions;
- Evaluation of the best type of application;
- Experimental determination of missing physical values of candidate
  products;
- If necessary, formulation of suitable products;
- Manufacture sample lots of products to be tested;
- Thermal cycling of samples and observation of variations in properties.

../..

Total cost:
     FF.   512,160
     u.a.   92,215
	E.C. Contribution: 36%
	     FF. 221,253
	     u.a. 39,837

COMMISSION OF THE
EUROPEAN COMMUNITIES

ENERGY R & D PROGRAMME

Objective: Solar Energy

Project or Sector:

Solar heat collectors and their
application to dwellings

3. Status

After six months' work.

As it was not possible to obtain samples or recipes of formulated
products from outside sources, a research had to be undertaken to
make such products.   The temperature range aimed at is 30-50°C.

A cycling apparatus of 50 liters capacity was built.

Several types of plastic capsules were selected on the basis of
possible mass production.

No actual results available yet.

COMMISSION OF THE EUROPEAN COMMUNITIES ENERGY R & D PROGRAMME Objective: Solar Energy	Project or Sector: Solar heat collectors and their application to dwellings

Title:

Study of a solar heating system with centralized production and storage, suited to grouped houses.

Duration: 12 months Period: 1/10/76 - 30/10/77	Contract No: 171-76 ESF Project No:

Contractor: Roulier - Costic - C. E. A.

Address: Présenté-Roulier, 54 Rue Jenner, 75013 Paris, France

Head of Project: M. Présenté

Description of research work

I. Objectives (aims)

Study of a solar heating system including long duration storage collector 'plant' and centralized auxiliary energy.   Precise simulation of the whole system (collection, storage, distribution, needs, ...).

2. Work Programme

1st six months:

- Architectural and thermal study of the buildings themselves;
- Study and development of simulation codes allowing a technical and an economic optimization of the whole system.

2nd six months:

- Analysis and improvement of the results given by the 1st six month's study in order to achieve plans and start the construction of the first 100 apartments at the end of 1977.

3. Status

Work in progress.

Total cost:
    FF.   512,658
    u.a.   92,304
	E.C. Contribution:
	    FF. 166,421
	    u.a. 29,964

Project B

Self-contained generating sets for the
production of mechanical and/or electrical power

COMMISSION OF THE EUROPEAN COMMUNITIES ENERGY R & D PROGRAMME Objective: Solar Energy	Project or Sector: Self-contained generating sets for the production of mechanical and for electrical power

Title:

      Self-contained generating sets for the production of electrical power.

Duration: 6 months Period: 1/5/76 - 31/10/76	Contract No: 032-76 ESD Project No:

Contractor: Messerschmitt-Bölkow-Blohm GmbH

Address: D-8000 München 80 - Postfach 801169 - Germany

Head of Project: Mr. H. Hopmann

Description of research work

I. Objectives (aims)

The aim of the study is to investigate major aspects concerning the technical feasibility and system definition for a helioelectric power plant of the multiheliostat/central receiver type. Whereas other contractors of the EEC do study other aspects of such a helioelectric power plant, MBB and its co-operating firm ANSALDO (Genova, Italy) concentrate on the following items:

- concentration system (heliostat, tracking system))
- civil engineering) MBB
- electrical system and interface to grid)
- receiver)
- steam cycle and control) ANSALDO
- turbine)
- interface with storage)

Together with the co-ordinator and other contractors of the EEC, MMB and ANSALDO shall participate in the definition of the overall system of a helioelectric power plant producing electric power of 1 MW. This will include the preparation of a practical programme plan and the order of magnitude cost estimates for the construction of a demonstration plant.

Total cost:
  DM. 438,000
  u.a. 119,672
	E.C. Contribution: 50%
	  DM. 219,000
	  u.a. 59,836

COMMISSION OF THE EUROPEAN COMMUNITIES ENERGY R & D PROGRAMME Objective: Solar Energy	Project or Sector: Self-contained generating sets for the production of mechanical and for electrical power

2. Work Programme

The concentrating system and the receiver i.e. the systems which are new elements in the design of power plants and which are the essential characteristics in an heliostat/central receiver solar power plant, were analysed in a systematic manner. For the concentrating system this represented a trade-off of several heliostat designs; the receiver will be from the type designed by Prof. Francia, Genova, and has been investigated by him for the special application in a 1 MW_{el} - plant.

Subsystems of a more conventional design, such as steam cycle and electrical system can be dealt with in a straight forward manner, keeping in mind the special requirement of a solar power plant. Interfaces with the heat storage system and special aspects of the heliostat field layout had to be discussed with other contractors of the EEC and the co-ordinator in the various expert meetings.

Preparation of programme plan and cost estimation for a 1 MW_{el} pilot plant was performed after the technical concept has been established in the expert meetings. The process of co-ordinating the various technical aspects has to be continued in the forthcoming project phases in order to implement a pilot plant which is optimised with respect to system performance and gives low investment cost for later commercial plants.

3. Status

For the subsystems listed in (1), a technical concept has been established which, after an optimisation with respect to system performance and after preparation of a detailed design, can be implemented to construct a 1 MW_{el} helioelectric demonstration plant. It is estimated that the implementation phase (development phase - phase B - plus production and installation phase - phase C -) will require 3 years.

The cost estimates are based on the technical concept and on the time schedule. As the site was not defined, no costs for civil engineering activities could be included.

The study has been completed and is delivered both to EEC and to the co-ordinator for incorporation into an overall system study.

COMMISSION OF THE EUROPEAN COMMUNITIES ENERGY R & D PROGRAMME Objective: Solar Energy	Project or Sector: Self-contained generating sets for the production of mechanical and/or electrical power

Title:

Feasibility study of a 1 MW(e) solar power generating system and preparation of a practical programme plan.

Duration: 10½ months Period: 15/2/76 - 31/12/76	Contract No: 044-76 ESUK Project No:

Contractor: General Technology Systems Limited (G.T.S.)

Address: 8, St. Bride Street, London EC4, United Kingdom

Head of Project: Dr. W.H. Stephens

Description of research work

I. Objectives (aims)

The primary objective is to co-ordinate on behalf of the Commission the work of expert groups at Contractors in participating countries necessary to establish the technical feasibility and order of magnitude cost estimates of a 1 MW(el) solar thermal electric generating system based on the central receiver/heliostat field principle. This includes particularly the analysis of proposed design configurations in order to optimise the overall system concept and ensure technical compatibility between subsystem and engineering interfaces, together with the preparation of a programme plan for development of the later phases of the Project, based on co-ordination of information from contractors.

2. Work Programme

Specialist discussions with experts in CNRS and EdF (France), MBB (Germany), Ansaldo (Italy) and JRC (Ispra) to agree the main design parameters for the demonstration plant and prepare the work programme, timescale and cost plan for phase B. Areas covered include definition of layout of the heliostat field; choice and preliminary design of heliostats and tracking system, thermal storage concept, receiver, steam cycle and electrical generator.

3. Status

Final report: "1 MW(el) Helioelectric Demonstration Generating Plant - Design Concept, Technical Feasibility and System Definition" No. GT 75085/3 delivered November, 1976.
A summary report No. GT 75085/4 was also delivered in December, 1976.

Total cost: £ 16,620 u.a. 39,888	E.C. Contribution: 100% £ 16,620 u.a. 39,888

COMMISSION OF THE EUROPEAN COMMUNITIES ENERGY R & D PROGRAMME Objective: Solar Energy	Project or Sector: Self-contained generating sets for the production of mechanical and/or electrical power

Title:

Use of solar heat for the production of power.

Duration: 6 months Period: 1/7/76 - 31/12/76	Contract No: 172-76 ESF Project No:

Contractor: Centre National pour la Recherche Scientifique (CNRS)

Address: LAAS, 7 Avenue du Colonel Roche, 31400 Toulouse, France

Head of Project: M. J.L. Abatut

Description of research work

I. Objectives (aims)

(a) The definition of a reference configuration for a helio-thermal electric power station rated at 1 MWe of the heliostat field-tower. type, in cooperation with the other contractors (MBB, ANSALDO, GTSL);

(b) The definition of a short-term (0.5-2 h) storage system designed to offset relatively short periods of overcast cloud conditions (problem dealt with by EdF);

(c) Synthesis of the heliostat field on the basis of a simulation analysis of the heliostat field, with an analysis of heliostat costs, including the control system.

2. Work Programme

- Analysis of the various thermal storage processes;

- Definition of a parallel storage system using molten salts; calculated dimensions;

- Simulation of the heliostat field. Analysis of shadowing and cosine factors;

- Optimum position of heliostats on a given site. Analysis of the influence of cloud-cover rate at ground level;

Total cost:
   FF.   166,620
   u.a.   30,000
	E.C. Contribution: 100%
	   FF. 166,620
	   u.a. 30,000

COMMISSION OF THE EUROPEAN COMMUNITIES ENERGY R & D PROGRAMME Objective: Solar Energy	Project or Sector: Self-contained generating sets for the production of mechanical and/or electrical power

2. Work Programme cont/..

- Evaluation of costs per square metre of heliostat (work subcontracted to industrial firm), including the guidance system.

3. Status

All the work on the foregoing programme has been completed:

(a) storage;

(b) positioning of heliostats;

(c) cost of a heliostat.

COMMISSION OF THE EUROPEAN COMMUNITIES ENERGY R & D PROGRAMME Objective: Solar Energy	Project or Sector: Self-contained generating sets for the production of mechanical and/or electrical power

Title:

       Organizational and managerial aspects of phase B (1MWe solar power generating system)

Duration: 4 months Period: 1/12/76 - 31/3/77	Contract No: 205-77 ESUK Project No:

Contractor: General Technology System Limited (G.T.S.)

Address: 8, St. Bride Street, London EC4, United Kingdom

Head of Project: Dr. W.H. Stephens

Description of research work

I. Objectives (aims)

In order to ensure the satisfactory completion of the overall system concept definition, to finalise the engineering design specifications for the 1 MW(el) demonstration plant and to implement manufacture and installation on site of the complete prototype system, an effective project management system is required.   The objective of the work under this contract is the preparation of a management plan which will provide the Commission with all necessary procedures and information needed to control the systematic planning and monitoring of all activities undertaken by the Contractors, consistent with financial and timescale constraints.

2. Work Programme

Study possible forms of organisation and planning procedures compatible with the interests of the Commission and participating Member States in order to identify a management system capable of giving effective control of all tasks undertaken by individual Contractors in phases B & C of the Project in accordance with the overall system concept resulting from Phase A.   The study includes recommendation of procedures for definition of the overall system specification and individual task specifications as the basis for contract actions, together with action necessary to ensure efficient co-ordination of the programme and provision of regular information to permit satisfactory monitoring of work progress and expenditure, with design reviews at appropriate times.   A committee structure to be operated by DG XII is proposed for control of technical and policy aspects of the overall programme.

Total cost:
    £   3,000
    u.a. 7,200
	E.C. Contribution: 100%
	    £ 3,000
	    u.a. 7,200

COMMISSION OF THE EUROPEAN COMMUNITIES ENERGY R & D PROGRAMME Objective: Solar Energy	Project or Sector: Self-contained generating sets for the production of mechanical and/or electrical power

3. Status

A report "EEC Solar Energy Project B 1 MW(el) Helioelectric Demonstration Generating Plant - Proposed Management System" GT 75085/5 was delivered in March 1977.

Project C

Photovoltaic conversion

COMMISSION OF THE EUROPEAN COMMUNITIES ENERGY R & D PROGRAMME Objective: Solar Energy	Project or Sector: Photovoltaic conversion

| Title:

    Concentrating solar systems for photovoltaic conversion. ||

Duration: 12 months Period: 1/6/76 - 31/5/77	Contract No: 104-76 ESD Project No: C/D/093(D)

Contractor: Forschungs-und Entwicklungslaber Kleinwächter

Address: D-7850 Lörrach, Germany

Head of Project: Prof. Dr-Ing. H. Kleinwächter

Description of research work

I. Objectives (aims)

The objective of the research work is, to develop low prize solar concentrators by using non-conventional technologies. Based on more than one year of pre-work of the contractor, prototypes will be built and combined with different photovoltaic cells.

Especially the inner mirrored funnel seems to be well adapted to solar cells, because it produces on a circular plan a concentration with homogeneous distribution of the energy density.

Different types of concentrators will be studied in theory and practice, especially under the point of view of low prize realisation of the installation as well in the countries of the E.C. as in developing countries with hard environmental conditions (sand, wind a.s.o.).

The whole system shall function during years without great service. Producing electricity and hot water or steam at the same time, such a combined system can be called hybrid collector.

2. Work Programme

(a) Definition of the principal technical parameters, such as:

- concentration ratio;
- power-range;
- availability;
- environment;
- need of surface;   ../..

Total cost:
    DM.   170,950
    u.a.   46,708
	E.C. Contribution: 75%
	    DM. 128,213
	    u.a. 35,031

COMMISSION OF THE EUROPEAN COMMUNITIES ENERGY R & D PROGRAMME Objective: Solar Energy	Project or Sector: Photovoltaic conversion

2. Work Programme cont/..

(b) Focusing systems:

- inner mirrored funnel;
- with flat plate mirrors;
- with paraboloid of rotation;
- with cylindrical paraboloid;

(c) Tracking mechanism:

- systems without tracking;
- systems with tracking;

(d) Cooling systems:

- heat exchanger;
- regenerative cooling;

(e) Construction and test of some prototypes.

3. Status

The points a – c of the working programme have been accomplished.
Thermal measurements of the concentrators are starting. We are
waiting for different samples of photovoltaic cells which should be
sent to us by some contractors of the project C, following the decision
during the meeting of March '77 in Brussels.

COMMISSION OF THE
EUROPEAN COMMUNITIES

ENERGY R & D PROGRAMME

Objective: Solar Energy

Project or Sector:

Photovoltaic conversion

Title:

Study of processes for reducing solar cell manufacturing and interconnection costs.

Duration: 18 months

Period: 1/9/76 - 28/2/78

Contract No: 106-76 ESF

Project No: C/A/058(F)

Contractor: R.T.C. La Radiotechnique - Compelec

Address: 51, rue Carnot, 92 Suresnes, France

Head of Project: Y. Salles

Description of research work

I.   Objectives (aims)

The aim of the research is to develop manufacturing and interconnection processes that would be cheaper and suitable for mass production with automated methods.

2.   Work Programme

Extension of present processes to the production of large cells in view of analysing the performances obtained when using well established production techniques.

Development of cell production with new processes and preliminary studies of interconnection technology.

3.   Status

Work in progress.

Total cost:
     FF.   726,749
     u.a.   130,847

E.C. Contribution: 54%
     FF.   470,933
     u.a.   84,789

COMMISSION OF THE EUROPEAN COMMUNITIES ENERGY R & D PROGRAMME **Objective:** Solar Energy	**Project or Sector:** Photovoltaic conversion

Title:

    Industrial technology for low cost high quality encapsulation of large solar cell arrays.

Duration: 18 months **Period:** 1/9/76 - 31/2/78	**Contract No:** 107-76 ESF **Project No:** C/F/031(F)

Contractor: R.T.C. La Radiothechnique Compelec

Address: 51, rue Carnot, 92 Suresnes, France

Head of Project: Mr. Y. Salles

Description of research work

I. Objectives (aims)

The proposed research aims at adapting the sandwich glass techniques, already used in safety windscreen or in buildings, to solar cell encapsulation. The use of this solution for solar cell encapsulation is original. It should provide solar cell arrays with an exceptional and well established weather resistance, cut the cost per m² and thus make possible the manufacture of self sustaining large dimension arrays which could easily fit into building infrastructures.

2. Work Programme

This programme is divided into three main steps:

1. A feasibility study.
   During this phase, the following items will be determined, based on small samples of 3 or 4 cells connected in series:

   - values of the main parameters: pressure, temperature, assembly method, which should allow the insertion of solar cells between two sheets of glass without mechanical breakage nor modification of their electrical properties;

   - means of attachment of the electrical terminals;

   - means of attachment of the arrays on a fixture.

          ../..

Total cost:
   FF. 1,504,727
   u.a. 270,917
	E.C. Contribution: 29.5%
	   FF. 532,673
	   u.a. 95,908

COMMISSION OF THE EUROPEAN COMMUNITIES ENERGY R & D PROGRAMME Objective: Solar Energy	Project or Sector: Photovoltaic conversion

2. Work Programme cont/..

B. A prototype design.
The experience gained during the feasibility study shall be extrapolated to the actual dimensions of a solar cell array such as it is foreseen.   New problems caused by the change of scale will need to be solved, i.e.:

- uniformity and homogeneity of pressures and temperatures during the sandwich process;

- mechanical stresses generated by expansion differences in large size solar cells;

- interconnections between a number of series mounted cells on which high pressures are exerted in low viscosity medium;

- development of the array attachment techniques.

C. A reproducibility and climatic tests study.
After step B, the manufacturing procedures shall be experienced again on a number of specimens in view of investigating the various parameters.

These specimens will be submitted to climatic tests and reliability assessment according to the methods which have already been developed by RTC for its previous solar cell arrays.

3. Status

Work in progress.

COMMISSION OF THE
EUROPEAN COMMUNITIES

ENERGY R & D PROGRAMME

Objective: Solar Energy

Project or Sector:

Photovoltaic conversion

Title:

Optimization of silicon solar cells intended to be
operated under medium solar concentration (10 to 50).

Duration: 12 months

Period: 1/7/76 - 30/6/77

Contract No: 109-76 ESF

Project No: C/A/020(F)

Contractor: Laboratoires d'Electronique et de Physique Appliquée (LEP)

Address: 3, Avenue Descartes, 94 Limeil-Brevannes, France

Head of Project: Emmanuel Fabre

Description of research work

I. Objectives (aims)

The proposed research is aimed towards an optimization of single
crystal silicon solar cells for   efficient operation under medium
range concentration (10 to 50 suns).

This optimization will be achieved through an important decrease
of the series resistance of the cell, by a factor 10 or more.

2. Work Programme

- Contribution of the N^+ diffused layer to the series resistance:
it will be decreased by using (i) a fine grid pattern as the front
contact and (ii) various diffusion conditions;

- Contribution of the base material resistivity: most of the cells
will be made from 1Ω. cm, p- type silicon, but lower resistivity
material will also be used for some experiments;

- The possible use of highly conductive antireflective coatings will
be investigated for a further decrease of the series resistance.

../..

Total cost:

    FF.   591,173
    u.a.   106,437

E.C. Contribution: 31%

    FF.   219,917
    u.a.   39,595

COMMISSION OF THE EUROPEAN COMMUNITIES ENERGY R & D PROGRAMME Objective: Solar Energy	Project or Sector: Photovoltaic conversion

3. Status

We have first solved the problem of metal-silicon rear contact resistance which, in fact, was giving the major contribution to the series resistance.   The use of Ti-Ag onto p- type silicon is prohibited, but Al-Ag onto p- type silicon or Ti-Ag onto p$^+$ -type silicon are found to be quite suitable.

The use of a fine grid pattern as the front metallic contact has already led to a substantial decrease of the series resistance, from 0.35Ω ("standard" cells) down to 0.05Ω.   We are at present carrying out a detailed analysis of the influence of the fine grid geometry on the series resistance, and the effect of the junction depth.   This work is done in close cooperation with Leuven Catholic University, where a computer simulation has been developed.   The effect of highly conductive antireflective coating and base material resistivity will be investigated in the last quarter.

COMMISSION OF THE EUROPEAN COMMUNITIES ENERGY R & D PROGRAMME Objective: Solar Energy	Project or Sector: Photovoltaic conversion

Title:

Deposition of polycrystalline silicon layers from the melt on conductive substrates.

Duration: 12 months Period: 1/7/76 - 30/6/77	Contract No: Project No:	110-76 ESF C/B/024(F)

Contractor: Laboratoires d'Electronique et de Physique appliquée (LEP)

Address: 3, Avenue Descartes, 94 Limeil Brevannes, France

Head of Project: Mr. Belouet

Description of research work

I. Objectives (aims)

The objective of this research is to develop a method for the deposition of polycrystalline silicon layers from the melt on ribbon-like carbon substrates, in order to achieve low cost solar cells for terrestrial applications.   In this deposition method, the carbon ribbons are pulled upwards against the molten top of a polycrystalline silicon pedestal. During this process, each ribbon draws along a silicon film which upon freezing gives the polycrystalline layer.

2. Work Programme

This project is essentially split into three main activities:

(a) Design and improvement of the pulling apparatus;

(b) Development and testing of appropriate carbon ribbons in cooperation with Le Carbone Lorraine - France - and the Philips Research Laboratories - Aachen (Germany);

(c) Characterization by physical and chemical means of the grown layers. Besides our own efforts, cooperation has been set up with C.N.R.S. (Toulouse) for the observation of crystal defects by the transmission electron microscopy technique, CEST (Toulouse) for photoconductivity transient measurements and the Philips Research Laboratories (The Netherlands) for chemical analyses.

..../..

Total cost:
    FF.   1,200,136
    u.a.   216,077
	E.C. Contribution: 39%
	    FF. 561,664
	    u.a. 101,124

COMMISSION OF THE EUROPEAN COMMUNITIES ENERGY R & D PROGRAMME Objective: Solar Energy	Project or Sector: Photovoltaic conversion

3. Status

At present, the first layers have been successfully pulled over lengths up to 15 cm. The deposition length happens to be limited by the insufficient melting of the top of the pedestal. Very recently, this limitation was overcome by placing appropriate thermal valves in the fields of the R.F. heating coil.

Regarding the ribbon-like substrate, a structure has been selected which is made of a soft carbon base coated by a thin pyrocarbon layer. The base consists of rolled natural graphite which ensures the adequate mechanical properties of the ribbon, whereas the pyrocarbon coating minimizes the silicon carbon chemical reaction.

The layers grown on such substrates consist of large grains elongated along the temperature gradient (typical size, width: 100 to 500 μm; length: 300 to 2000 μm). The preferred surface orientations are (211) and (101) and the major defects are grain boundaries and twins. The resistivity of the layers thus prepared (p-type) is close to that of the starting material and the diffusion length of the minority carriers is always in excess of 20 μm ($\simeq$ 1Ω.cm). Finally, preliminary results on cell structures prepared from these layers have shown overall A M_1 conversion efficiencies around 5%.

The work in progress is specially centered on the perfecting of the R.F. furnace and the improvement of the ribbon substrate.

COMMISSION OF THE EUROPEAN COMMUNITIES ENERGY R & D PROGRAMME Objective: Solar Energy	Project or Sector: Photovoltaic conversion

Title:
 	
    Growth of silicon layers from vapour phase on a
    liquid metal layer.	

Duration: 18 months Period: 1/8/76 - 31/1/78	Contract No: 111-76 ESN Project No: C/B/089(N)

Contractor: Katholicke Universiteit Nijmegen Address: Toernooiveld, Nymegen, Netherlands Head of Project: Prof. Dr. J. Bloem	

Description of research work

  I. Objectives (aims)

The objective of the research is to obtain device quality silicon layers on a cheap substrate.   The chemical vapour deposition of silicon on a substrate which is covered with a fluid layer to enhance the crystallinity could give the desired results.

2. Work Programme

CVD (chemical vapour deposition) is carried out in a watercooled epitaxial reactor with SiH_4 as a source of silicon.   The addition of HCl to the gas phase can further be used to control the density of nuclei on the fluid film.   Stainless steel and graphite were selected as substrate material with tin as the preferred intermediate layer.

The programme comprises the study of:

(a) substrates (Fe and C)
(b) liquid layer (thickness, method of application)
(c) nucleation and growth of silicon
(d) optimisation of nucleation and growth
(e) doping
(f) evaluation.

                                              ../..

Total cost:
    Fl.   139,010
    u.a.   38,400
	E.C. Contribution: 72%
	    Fl. 100,000
	    u.a. 27,624

COMMISSION OF THE EUROPEAN COMMUNITIES ENERGY R & D PROGRAMME Objective: Solar Energy	Project or Sector: Photovoltaic conversion

3. Status

(a) C has been selected as preferable substrate material;

(b) A 5 µm thick Sn layer, doped with silicon is suitable;

(c) The addition of HCl reduces the number of nuclei so that each nucleus can grow out on the liquid tin layer.   Domains up to 250 µm have been obtained.

The work is now concentrated on point (d) of the programme;   (e) and (f) are under study in preliminary tests.

COMMISSION OF THE EUROPEAN COMMUNITIES ENERGY R & D PROGRAMME Objective: Solar Energy	Project or Sector: Photovoltaic conversion

Title:

        Gallium Arsenide Solar Cells.

Duration: 18 months Period: 1/7/76 - 31/12/77	Contract No: 113-76 ESUK Project No: C/A/059(UK)

Contractor: Plessey Company Ltd.

Address: Caswell, Towcester, Northants, United Kingdom

Head of Project: R. Davis

Description of research work

I. Objectives (aims)

To fabricate solar cells from multilayer $GaAs/Ga_{1-x}Al_xAs$ structures and assess their performance under concentrated solar power.

Epitaxial layers are to be grown on GaAs substrates using the metal alkyl, vapour phase method, this technique being the most suitable method of epitaxial growth for extension to large area slices. For example, solar cell structures may incorporate a diffused junction in GaAs and a window region of $Ga_{1-x}Al_xAs$ in which grading of the Al composition provides enhanced carrier collection. Standard methods of contacting grid pattern definition and packaging will be used, and considerable attention will be paid to the minimisation of series resistance.

2. Work Programme

(a) **Design Procedures:-** The influence of doping profile and grading of aluminium composition will be assessed using a theoretical model. Optimum layer configurations will be determined for operation of $Ga_{1-x}Al_xAs/GaAs$ cells at a variety of solar concentration levels.

(b) **Vapour Phase Epitaxy:-**

    (i)   The epitaxy kit designed for growth of p and n layers of GaAs and $Ga_{1-x}Al_xAs$ will be designed and constructed;

    (ii) Conditions for reproducible and uniform growth of epitaxial layers will be established;   ../..

Total cost:
  £   48,141
  u.a. 115,540
	E.C. Contribution: 49%
	  £ 23,750
	  u.a. 57,000

COMMISSION OF THE EUROPEAN COMMUNITIES ENERGY R & D PROGRAMME Objective: Solar Energy	Project or Sector: Photovoltaic conversion

2. Work Programme cont/...

(b) Vapour Phase Epitaxy:-

   (iii) Multilayer configurations will be grown to requirements for optimum solar cell performance.

(c) Liquid Phase Epitaxy:- Multilayer structures in $Ga_{1-x}Al_xAs/$ GaAs can be provided by our existing liquid phase epitaxy capability, and will be used in a complementary and comparative role, particularly before the VPE approach is established.

(d) Cell Fabrication and Assessment:-

   (i) Small test cells ($3 \times 3mm^2$) will be used initially to establish processing techniques, and establish correlation between materials parameters and cell performance;

   (ii) Existing device processing techniques will be adapted for contacting, pattern definition, chip separation, mounting and packaging. Grid pattern design and control of contact thickness by electroplating will be used to minimise series resistance. Anti-reflection coatings will be provided;

   (iii) Cell performance will be assessed in terms of I/V characteristics, series resistance, conversion efficiency and spectral response. Measurements will be carried out under normal and concentrated sunlight, and also under simulated solar conditions;

   (iv) Device area will be scaled up to 1 cm^2 and heat sinking designs developed for operation at power levels up to 1,000 suns;

   (v) Life tests on cells operating at room temperature will be started.

3. Status

The vapour phase epitaxial system has been constructed and recently commissioned. Preliminary layers of AlAs are limited to 2 μm in thickness and exhibit slow deterioration when exposed to air. Stable layers of GaAs with consistently good crystallographic appearance have been grown, and are being used for assessment of background impurity levels.

../..

COMMISSION OF THE EUROPEAN COMMUNITIES ENERGY R & D PROGRAMME Objective: Solar Energy	Project or Sector: Photovoltaic conversion

3. Status   cont/..

A theoretical model for graded band gap $Ga_{1-x}Al_xAs$/GaAs cells incorporating series resistance has been developed, and material requirements for optimum cell performance are being formulated. Using interim stocks of liquid phase epitaxial multilayer structures, the processing technology for 3 x 3 mm^2 cells has been established.   Measurements in sunlight and under simulated solar conditions have shown the cells to be 5-6% efficient at 80 mW/cm^2, and to be limited by low current levels due to the material structures not being optimum for solar cell applications.

Values of open-circuit voltage (0.95 - 1.00) and fill factor (0.75 - 0.80) are satisfactory.   The processing is currently being applied to more optimum material structures, and anti-reflection coatings introduced using anodisation of $Ga_{1-x}Al_xAs$.   A test rig has been designed for cell assessment, which includes a sun tracking facility and a concentrator system for up to 1,000 suns.

COMMISSION OF THE EUROPEAN COMMUNITIES ENERGY R & D PROGRAMME Objective: Solar Energy	Project or Sector: Photovoltaic conversion

Title: Thin film solar cells.	

Duration: 12 months Period: 1/9/76 - 31/8/77	Contract No: 120-76 ESUK Project No: C/A/060(UK)

Contractor: Plessey Company Ltd. Address: Caswell, Towchester, Northants, United Kingdom Head of Project: G..J. Rees	

Description of research work

I. Objectives (aims)

To study the factors at present limiting the performance of thin film solar cells and to determine the materials and device structures most likely to satisfy the requirements of efficiency and low cost.

2. Work Programme

(a) Conduct a literature survey to review the materials and technologies currently used in thin film solar cells and critically consider the factors limiting performance;

(b) Assess the advantages of various possible device structures with regard to material properties and their use in thin film structures;

(c) Consider the problem of substrate compatibility with materials, with the view of proposing further possible thin film devices;

(d) Consider the economics of thin film solar cells with regard to the cost and availability of materials, the production processes involved and device lifetime;

(e) Study the suitability of thin film photovoltaics with regard to the climate.

Calculations are to be performed wherever possible and necessary to further our understanding of the devices and to assess their future potential.

Total cost:
  £   14,500
  u.a.   34,800
	E.C. Contribution: 100%
	  £ 14,500
	  u.a. 34,800

COMMISSION OF THE EUROPEAN COMMUNITIES ENERGY R & D PROGRAMME Objective: Solar Energy	Project or Sector: Photovoltaic conversion

3. Status

Much of the materials and device survey (a) has been completed, including CdS/Cu$_2$S, CdTe, InP/CdS and Si, but is subject to continual updating.   Of these, amorphous silicon looks particularly promising in view of recent experimental results.   This material possesses an expanded optical gap and a large absorption coefficient, compared with crystalline silicon, and good carrier lifetimes.   Although mobilities initially appear too small to give diffusion lengths sufficiently long to enable good carrier collection efficiency, it appears that an adequate short circuit current can be achieved through the use of an appropriately designed structure.   Calculations are in progress to further our understanding of the physical processes involved in amorphous silicon devices and to determine the optimum device structure.

The review of device structures (b) is in progress and we shall be looking at the economic and climatic considerations shortly.

COMMISSION OF THE EUROPEAN COMMUNITIES ENERGY R & D PROGRAMME Objective: Solar Energy	Project or Sector: Photovoltaic conversion

| Title:

Development of a process for manufacturing silicon films used in solar cells by means of rheotaxy on tinned sheets. ||

Duration: 12 months Period: 1/9/76 - 31/8/77	Contract No: 125-76 ESF Project No: C/B/027(F)

| Contractor: CEA-CEN Grenoble (CENG), Laboratoire d'Electronique et de Technologie de l'Informatique

Address: rue de la Federation 29-33, 75015 Paris

Head of Project: M. Elston ||

Description of research work

I. Objectives (aims)

To find an economical film process technology for depositing a polycrystalline silicon layer with 50 μ size grains, and with a thickness of 50 μ.

These silicon films are to be used for photovoltaic solar cells. The originality of the research is the use of a thin tin film between the substrate and the silicon film deposited. During deposition the layer of fluid tin increases the growth of the silicon seeds. We have observed, besides, that tin does not modify the electrical properties of silicon.

2. Work Programme - 3. Status

During the first phase we tried to develop the rheotaxial process on steel substrates. But even with an iron barrier we noted that silicon layers were contaminated by iron and grain size was limited by the low temperature deposition.

Proposed work for next phase:

- Replacing the steel substrate with a graphite substrate;

- Realizing the silicon deposition in a low pressure CVD process;

- Investigating the effect of crystal growth conditions on the physical and electrical properties of the polycrystalline silicon layers.

Total cost: FF. 319,700 u.a. 57,560	E.C. Contribution: 60% FF. 191,820 u.a. 34,536

COMMISSION OF THE EUROPEAN COMMUNITIES ENERGY R & D PROGRAMME Objective: Solar Energy	Project or Sector: Photovoltaic conversion

Title:

Development of single crystal CdTe solar cells for terrestrial application suitable for use in optical concentrators (concentration ratio 50:1 and higher).

Duration: 18 months Period: 1/10/76 – 31/3/78	Contract No: 132–76 ESD Project No: C/A/086(D)

Contractor: Battelle-Institut e.V.

Address: D6 Frankfurt/M., Postfach 900160, Germany

Head of Project: Dr. Hans Jäger

Description of research work

I. Objectives (aims)

Solar cells to be used in optical concentrators with high concentration ratio (CR) are to be developed on the basis of CdTe crystals. The CdTe bandgap of 1.5 eV at 300 K is optimally fitted to the solar emission spectrum. As a semi-conductor material, CdTe of high purity can be obtained cheaply in quantity. To obtain highly conductive window layers and simultaneously photovoltaic cells the aim of the device development is an antipolar semi-conductor heterolayer of larger bandgap II-VI compounds such as CdS and ZnTe to be deposited on the CdTe substrate. To get contact and collector resistances down to very low values special geometric solutions are provided.

2. Work Programme

CdTe crystals are to be grown by the Bridgman method and with a high conductivity to be applied as substrates. Crystals of n-type as well as of p-type are planned to be grown. These crystals will be characterised with respect to electrical and structural properties. The next step will be the generation of a CdTe layer with low carrier density on such substrates. This will be done either by compensating diffusion or by sublimation of CdTe. This layer is provided to contain the space charge layer of the photovoltaic cell. The corresponding diode will be formed by a semiconductor heterolayer of opposite polarity and of a larger bandgap. On n-CdTe p-ZnTe is provided; p-CdTe will be covered by n-CdS. All parts of the cell, above all the contacts, have to be of very low resistance because of high current densities due to the optical concentrator.

../..

Total cost:
    DM. 454,113
    u.a. 124,075
	E.C. Contribution: 50%
	    DM. 227,057
	    u.a. 62,037

COMMISSION OF THE EUROPEAN COMMUNITIES ENERGY R & D PROGRAMME Objective: Solar Energy	Project or Sector: Photovoltaic conversion

3. Status (March '77)

Several n-type and p-type crystals have been grown.   Structural
properties have been analysed by etching.   In most cases mono-
crystals were found. Hall data reveal high conductivity in
the n-type CdTe, and sufficient conductivity in p-type CdTe.
Annealing was done with n-CdTe to achieve the low conductivity
layer.   This technique seems to be difficult.   To check the
quality of whole-surface Schottky contacts 100 Å semitransparent
Au layers covered with a thicker Au-collector comb were evaporated
on the n-substrate material (still without a low carrier density
layer).   I-V-characteristics and photovoltaic response are
encouraging even without this special layer necessary for fitting
space charge layer depth and absorption depth to get in optimal
conversion efficiency.   Series resistances were determined.
Calculations of the heterojunction layer resistance depending on
the form of the metallic reticular collector were performed;
suitable material was purchased.

COMMISSION OF THE EUROPEAN COMMUNITIES ENERGY R & D PROGRAMME Objective: Solar Energy	Project or Sector: Photovoltaic conversion

| Title:

    Development of Low Cost Cadmium Sulphide Sintered
    Ceramic Ribbon Solar Cells for Terrestrial Applications. ||

Duration: 18 months Period: 1/10/76 - 1/4/78	Contract No: 145-76 ESUK Project No: C/A/035(UK)

| Contractor: G. V. Planer Ltd.

Address: Windmill Road, Sunbury-on-Thames, Middlesex, UK

Head of Project: P. Norgate ||

Description of research work

I.   Objectives (aims)

The aim of the proposed research is to exploit the potentially low cost fabrication techniques offered by ceramic production technology, in conjunction with low cost CdS starting material prepared by chemical precipitation methods, to develop low cost CdS solar cell arrays for terrestrial use.

2.   Work Programme

(a)   Investigation of chemical precipitation techniques for production of doped CdS starting material having reproducibly controlled chemical, physical and electrical characteristics;

(b)   Optimization of existing single cell preparation, activation and contacting techniques employing precipitated doped CdS starting material prepared under heading (a).   During the work on optimization of production techniques, due regard will be paid to the use of methods and materials which are suitable for eventual incorporation into a large scale production process based on the fabrication of continuous ribbon or large sheet cell arrays;

(c)   Investigation into the continuous production of n-type sintered ceramic CdS ribbon or large sheet.

3.   Status

Work in progress.

Total cost:
    £   69,146
    u.a. 165,950
	E.C. Contribution: 46%
	    £ 31,807
	    u.a. 76,337

COMMISSION OF THE EUROPEAN COMMUNITIES ENERGY R & D PROGRAMME Objective: Solar Energy	Project or Sector: Photovoltaic conversion

Title:	The development of a screen printed contact technique, associated with conversion efficiency enhancements and other manufacturing process economies, to reduce the cost of silicon solar cells to below £10 per watt, in array form.

Duration: 18 months Period: 1/7/76 - 31/12/77	Contract No: 146-76 ESUK Project No: C/A/030(UK)

Contractor: Ferranti Ltd.

Address: Simonway, Wythenshawe, Manchester M22 5LA, United Kingdom

Head of Project: Mr. A. V. Whale

Description of research work

I. Objectives (aims)

The aim of the proposed research is to reduce the cost of silicon solar cell arrays to less than £10 per watt by reductions in processing costs, increases in the conversion efficiency (which is currently in the region of 9.5 to 10%), and assessment of new materials.

2. Work Programme

The following sectors are being investigated:

(a) development of a screen printed from contact;

A series of experiments is proposed to investigate the variation of contact quality with ink formulation on:

(i)   slices with optimum thickness of silicon dioxide grown during diffusion;

(ii)   slices with a more optimum anti-reflection coating such as titanium or tantalum oxides (see c below);

(iii) slices as in (i) and (ii) above but with a preferentially etched anti-reflection surface;

(iv)   slices as in (i), (ii), and (iii) above but with diffused junction depths in the region of 0.1 micron;

../..

Total cost: £ 72,975 u.a. 175,140	E. C. Contribution: 55% £ 39,771 u.a. 95,450

COMMISSION OF THE
EUROPEAN COMMUNITIES

ENERGY R & D PROGRAMME

Objective: Solar Energy

Project or Sector:

Photovoltaic conversion

2. Work Programme   cont/..

(b)   application of a screen printed back contact and an
investigation of the possibility of the simultaneous provision
of a back surface field;

(c)   application of more optimum anti-reflection coatings than the
thermal silicon dioxide established during the junction
diffusion;

(d)   provision of junction depths in a region of 0.1 micron.
Phosphorous oxychloride as a dopant source are to be used on
plain slices and on preferentially etched slices;

(e)   preferential front surface etching;

(f)   slice handling and production economies;

(g)   use of cheaper starting material.

3.   Status

Work in progress.

COMMISSION OF THE EUROPEAN COMMUNITIES ENERGY R & D PROGRAMME Objective: Solar Energy	Project or Sector: Photovoltaic conversion

Title:

Feasibility study of photovoltaic cells made from Silicon and GaAs to be used with concentrators.

Duration: 18 months Period: 1/7/76 - 31/12/77	Contract No: 150-76 ESI Project No: C/A/076(I)

Contractor: Laboratorio di Elettronica, Instituto di Fisica, Università di Modena

Address: Vivaldi, 70, Modena

Head of Project: Claudio Canali

Description of research work

I. Objectives (aims)

We propose a study of new techniques for the preparation of p-n or Schottky junctions, obtained at low temperatures (500°C) for silicon solar cells with particular attention to:- (a) the epitaxial growth and doping by solid phase of thin silicon layers and (b) the formation and the electrical characterization of silicon-silicides Schottky barriers obtained from interaction between silicon and thin metal films.

2. Work Programme

(a) Study of the epitaxial growth by solid phase of thin silicon layers (0.1-1 μm) using Si (crystal)/thin Pd film/amorphous silicon structures heated at temperatures below 500°C.   A check of different metal films (Al) with the evaluation of their role as diffusion and transport medium of amorphous silicon.

(b) Evaluation of the epitaxial growth, by solid phase, process: kinetics, transport rate of amorphous silicon, activation energy of regrowth process, effect of impurities (C, O), influence of substrate orientation.   Evaluation of possibility of the eteroepitaxial growth by solid phase of silicon on different substrate (sapphire).   The evaluation of the crystalline structure of the epitaxially grown layer will be partially performed at the "Donegani" Institute, Novara.   The study of the influence of the surface conditions on solid-phase epitaxy will also be studied in collaboration with F.O.M. Institute of Amsterdam.

../..

Total cost:
    Lit.   91,500,000
    u.a.   146,400
	E.C. Contribution: 26%
	    Lit. 23,437,500
	    u.a. 37,500

COMMISSION OF THE EUROPEAN COMMUNITIES ENERGY R & D PROGRAMME Objective: Solar Energy	Project or Sector: Photovoltaic conversion

2. Work Programme   cont/..

(c) Doping of the layer epitaxially grown, by solid phase, either
from inclusion of atoms of the metal film (i.e., Al) acting
as a dissolution and transport medium for amorphous silicon
or by adding impurities which behave as donors or acceptors.

(d) Electrical characterization of the silicon epitaxial layers
grown by solid phase and obtained p-n junctions.

(e) Study of the feasibility of thin, 100-200 Å, silicide films
of Pd, Pt by interaction between crystal silicon and metal
films at $T \leq 500°C$.   Evaluation of the lateral uniformity
and growth kinetics of these thin silicide films.

(f) Measurements of barrier height and electrical properties of
contacts between thin silicide film ($PtSi$, Pd_2Si) and n-type
silicon.

(g) Evaluation, with the cooperation of LAMEL laboratory (Bologna)
of solar cell prototypes with n-p and Schottky junctions
obtained by solid phase epitaxy and silicon-silicide contacts
respectively.

3. Status

(i)   Solid phase epitaxy can be obtained on large areas in systems
$Si(xtl)/Al/Si(a)$ and $Si(xtl)/Pd/Si(a)$;

(ii)   Impurities (Al, Sb) can be included in grown layers up to
solid solubility;

(iii) Thin PtSi layers of 100Å show a $\rho_s = 45 \mu\Omega cm$ ($S_{ptsi} = 30\ \mu\Omega cm$)
and $\rho_{s\square} = 45\ \Omega$.

COMMISSION OF THE EUROPEAN COMMUNITIES ENERGY R & D PROGRAMME Objective: Solar Energy	Project or Sector: Photovoltaic conversion

Title:

Optimization of silicon solar cells intended to be operated under medium solar concentration (10 to 50).

Duration: 18 months Period: 1/7/76 - 31/12/77	Contract No: 152-76 ESB Project No: C/A/105(B)

Contractor: Katholicke Universiteit Leuven (K.U.L.)

Address: Laboratorium ESAT - Kard.Mercierlaan 94 - 3030 Heverlee, Belgium

Head of Project: Prof. R. Van Overstraeten

Description of research work

I. Objectives (aims)

The objective of the programme is the optimization of the metal grid of silicon solar cells such that they can be used with high efficiency at illumination levels of 50 suns.   It has indeed been demonstrated that, using a one dimensional system, an average concentration with conventional silicon solar cells yields an economic improvement, provided that the series resistance of the cell can be made smaller.   The present day silicon solar cells are optimized for a one sun operation.   Their series resistance causes a voltage drop of about 35 mV at AM1.   This voltage drop would become larger than 300 mV at an illumination of 10 suns, such that the efficiency would drop by a factor of 2.

The aim of this study is to optimize silicon solar cells, using theoretical computer calculations, such that they can be used at illumination levels between 10 and 50 suns.   The results will be extensively compared with experimental data obtained by LEP, France.

2. Work Programme

In the first stage the series resistance will be calculated, using a two-dimensional programme.   This computer programme discretizes the solar cell in two dimensions, substitutes the lumps by electrical equivalents and finally solves the electrical network for its terminal characteristics from which the series resistance will be obtained. The development of such an analysis tool can be considered as the first step in the programme.

../..

Total cost:
    FB.   2,672,000
    u.a.   53,440
	E.C. Contribution: 46%
	    FB. 1,222,500
	    u.a. 24,450

COMMISSION OF THE
EUROPEAN COMMUNITIES

ENERGY R & D PROGRAMME

Objective: Solar Energy

Project or Sector:

Photovoltaic conversion

2. Work Programme cont/..

In the next stage the computer programme will be used to simulate
available solar cells and the computed values for the series
resistance will be compared with experimental data.   This stage
demonstrates the feasibility of the theoretical approach.

The last stage is a detailed optimization of the efficiency at
various suns, for a variety of grid patterns.   Graphs showing
optimum grid characteristics for every sun value must be set up.
In addition, detailed information about the different posts of the
series resistance must be given for every case.

3. Status

The first part of the programme, the development of a computer
analysis tool to calculate series resistance of solar cells in a
two dimensional way has been preformed.   The computer programme
simulates in an efficient way, the IV characteristics of solar cell
with up to 30 metal fingers.

The results of the computer programme have been extensively compared
with the experimental data obtained by LEP.   A good agreement is
found, provided that the contact resistance problem in the cells is
eliminated.   This was actually achieved for cells with a back
surface field diffusion.   For cells without back surface field
diffusion the series resistance is contact limited and there is no
correlation with the computed results.

The computations also indicated that operation at 50 suns will be
possible.

Optimization of the structures for different concentrations will be
done in the next period, together with an analysis of rectangular
and hexagonal structures.

COMMISSION OF THE EUROPEAN COMMUNITIES ENERGY R & D PROGRAMME Objective: Solar Energy	Project or Sector: Photovoltaic conversion

Title:
Theoretical study of alternative thin film solar cells, and experimental realisation of the InP-CdS thin film solar cell.

Duration: 18 months Period: 1/9/76 - 31/2/77	Contract No: 154-76 ESB Project No: C/A/090(B)

Contractor: Universiteitsvermogen Rijksuniversiteit Gent

Address: De Rijksuniversiteit Gent,
St Pietersniewstraat, 42900, Gent, Belgium

Head of Project: Prof. H. Pauwels

Description of research work

I. Objectives (aims)

The proposed research envisions on the one hand to establish on theoretical grounds criteria which allow a justified choice among the many possible combinations of materials. On the other hand, the experimental realisation of a potentially promising alternative combination, i.e. the InP-CdS solar cell in purely thin film technology, is envisioned.

2. Work Programme

In the theoretical part, criteria will be established which allow to choose the most appropriate material combination for thin film solar cells which realize the best efficiency. The study of the influence of the energy gaps in heterojunctions were completed by co-workers of the laboratory (see section C). Important problems remain to be investigated: in heterojunctions energy barriers occur; the different materials consituting the heterojunction cause reflections of the incident light; surface states are present at the junction. Many material parameters play a part, but several of them are correlated: it is necessary to establish the minimum essential information which must be known about a material. These problems will also be investigated for metal-semiconductor structures (Schottky barriers). Once these criteria are known, it will be possible to determine by a systematic study of the literature the combination of materials for which the efficiency can be predicted, and the materials for which essential information is not available. Finally, measurement procedures to determine this missing information will be investigated.

../..

Total cost:
    FB. 3,102,248
    u.a. 62,045
	E.C. Contribution: 28%
	    FB. 867,500
	    u.a. 17,350

<table>
<tr><td>COMMISSION OF THE
EUROPEAN COMMUNITIES

ENERGY R & D PROGRAMME

Objective: Solar Energy</td><td>Project or Sector:

Photovoltaic conversion</td></tr>
</table>

2. <u>Work Programme</u> cont/..

In the experimental part, the InP-CdS solar cell will be realized
in purely thin film technology. From the many possibilities this
choice has been made because of the potentially large efficiency
(12.5%) and because of the experience of our research group with
CdS and with InSb.   The high prize of
InP-powders makes the use of the two-source method essential, i.e.,
the starting materials will be Indium and Phosphor. The first
phase of the research will therefore consist in trying to evaporate
InP with this method, and the investigation of the obtained thin film
layer (determination of the doping, mobility, and diffusion length
of the minority carriers). Next, the junction InP-CdS will be
investigated: diode characteristics, lifetime of the minority
carriers, energy barriers, surface states, behaviour as a solar cell.
Finally, this information will allow to build a model of the InP-CdS
solar cell.

3. <u>Status</u>

Work in progress.

COMMISSION OF THE EUROPEAN COMMUNITIES ENERGY R & D PROGRAMME Objective: Solar Energy	Project or Sector: Photovoltaic conversion

Title:

(Re)crystallization of amorphous and small grain size polycrystalline silicon.

Duration: 18 months Period: 1/10/76 - 31/3/78	Contract No: 155-76 ESN Project No: C/B/094(N)

Contractor: Rijksuniversiteit Utrecht

Address: Sorbonnelaan 4, Netherlands

Head of Project: Prof. Dr. Ir. S. Radelaar

Description of research work

I. Objectives (aims)

(a) To obtain sufficient information to control the recrystallization of fine-grained polycrystalline Si by uniform heating methods;

(b) To gain information on the recrystallization behaviour of the fine-grained polycrystalline Si in temperature gradients and (or) controlled atmospheres;

(c) To characterize the obtained coarse-grained material and to study the quality of the material for solar cell production

2. Work Programme

(i) Study of the influence of temperature on recrystallization (Specimens heated in a furnace, or by RF-heating);

(ii) Influence of specimen thickness on the recrystallization behaviour (ibid.);

(iii) Investigation of the texture and defect structure of the obtained material;

(iv) A study of the influence of temperature gradients on recrystallization and grain growth;

(v) A determination of the influence of the surrounding gas atmosphere (vacuum, inert gases, reactive gases) on tertiary grain growth;   ../..

Total cost: Fl. 628,500 u.a. 173,619	E.C. Contribution: 25% Fl. 155,750 u.a. 43,025

COMMISSION OF THE EUROPEAN COMMUNITIES ENERGY R & D PROGRAMME Objective: Solar Energy	Project or Sector: Photovoltaic conversion

(vi)   A study of the influence of Al on grain growth;

(vii)   Measurements of the electrical parameters (resistivity, carrier mobility) of the coarse grained material;

(viii) Preliminary experiments on solar cells from recrystallization silicon;

(ix)   First experiments on the growth of amorphous and poly-crystalline silicon layers by means of chemical vapour deposition.

3. Status

Research on items (i), (iii) and (iv) is practically completed. Work on items (ii) and (vi) has recently been started.

COMMISSION OF THE EUROPEAN COMMUNITIES ENERGY R & D PROGRAMME Objective: Solar Energy	Project or Sector: Photovoltaic conversion

| Title:

Influence of surface structure and surface adsorbates on solid phase epitaxial growth. ||

Duration: 18 months Period: August 1976 - February 1978	Contract No: 156-76 ESN Project No: G/B/092(N)

Contractor: Stichting F.O.M.

Address: Lucasbolwerk 4, Utrecht, Netherlands

Head of Project: Prof. Dr. J. Kistemaker; Dr. F. W. Saris,   Fom-Instituut
voor Atoom- en Molecuulfysica. Kruislaan 407

Description of research work

I. Objectives (aims)

The production of thin layers of semi-conductor materials by solid phase epitaxial growth can lead to a substantial decrease in the costs of solar cell production. The purpose of the research is to study the influence of surface structures and surface adsorbates on solid phase epitaxy, in order to be able to indicate conditions (vacuum, temperature, materials) under which optimal results can be obtained.

2. Work Programme

We shall start investigating homo-epitaxy on Si under UHV conditions. A Si single crystalline surface will be sputter-cleaned. As a result of sputtering the surface layer will become amorphous. On top of this amorphous layer a new amorphous Si layer will be freshly evaporated.

Since we are able to monitor surface coverage of C and O in situ, we will be able to make sure that the Si-interface is contaminant free before Si evaporation. The thickness of the amourphous layer deposited will be measured in situ by proton backscattering and channeling which will also be used to determine whether epitaxial growth is established after heating the sample at a temperature of $500\text{-}600^\circ C$.

If this process is successful, we will try to grow doped homo-epitaxial layers in a similar way and we will also investigate epitaxial growth on less expensive Si crystals (ribbon).

../..

Total cost:
      Fl.   283,750
      u.a.   78,384
	E.C. Contribution: 48%
	      Fl. 135,870
	      u.a. 37,533

COMMISSION OF THE EUROPEAN COMMUNITIES ENERGY R & D PROGRAMME Objective: Solar Energy	Project or Sector: Photovoltaic conversion

3. Status (February 1977)

(a) The Rutherford Backscattering Technique has been demonstrated as a suitable technique to investigate Solid Phase Epitaxial Growth (SPEG);

(b) Experience has been obtained in depositing layers of silicon by e-gun evaporation sources in ultra high vacuum;

(c) Results already obtained indicate that the SPEG method works;

(d) We combined the Rutherford Backscattering Technique with Ultra-high Vacuum evaporation.

COMMISSION OF THE EUROPEAN COMMUNITIES ENERGY R & D PROGRAMME Objective: Solar Energy	Project or Sector: Photovoltaic conversion

Title:

Spray method for photovoltaic material preparation (principal aim: CdS photocells).

Duration: 18 months Period: 1/7/76 - 31/12/77	Contract No: 159-76 ESF Project No: C/A/023(F)

Contractor: —Université des Sciences et Techniques du Languedoc, Montpellier;
—Université du Haut-Rhin, Mulhouse;
—Ecole Nationale Supérieure de Chimie de Paris.

Address: Université des Sciences et Techniques du Languedoc, 34060 Montpellier, France

Head of Project: Prof. Michel Savelli

Description of research work

I. Objectives (aims)

The aim is the preparation of metallic and semiconducting materials of CdS-Cu$_2$S photocells by spray on glass.

2. Work Programme

The following sectors are to be investigated:

A. Research of convenient solving-vector for sprayed Cu$_2$S films

- Investigation of formation and stability of "copper-thiourée" compound and eventually thioacétamide, thiosemicarbazide in acétonitrile;

- Investigation of thiourée oxidation by copper (11) and influence of organic solvant;

- Preparation of copper (1)-thiourée compound and investigation of their thermal decomposition; chemical and electrochemical analysis of decomposition products; comportment of these compounds in water or other organic solvants.

B. Physicochemical investigation of spray with experimental nozzle

- Realisation of experimental nozzle;

- Test of solutions comportment;

../..

Total cost:
    FF.   422,000
    u.a.   75,981
	E. C. Contribution: 50%
	    FF. 211,000
	    u.a. 37,991

COMMISSION OF THE
EUROPEAN COMMUNITIES

ENERGY R & D PROGRAMME

Objective: Solar Energy

Project or Sector:

Photovoltaic conversion

2. Work Programme   cont/..

- Control of aerosol dispersion depending on films physical properties;

- Research of different parameters controlling spray;

- Study of solvent vapors condensation and its recuperation;

- Security conditions due to the use of organic solvents.

C. Formation and characterization of Cds, Cu_2S, SnO_2, thin films on different support

- Realization of pulverisers specific for each material (area 10 cm x 10 cm);

- Study of nature and temperature influence of support on the reaction, the quality of film and its sticking on support;

- Physicochemical analysis of films;

- Physical characterization of films : electrical and optical properties.

D. Realization of CdS photocells or glass

- Research of adequate parameters of films fabrication;

- Characterization of photovoltaic properties of fabricated photocells.

Note: The participation of the three laboratories is as follows:

	1	2	3	4
L. E. A. A.	*	**	**	**
C. R. P. C. S. S.	**	*	**	**
G. E. M. C. E. S.		**	*	*

*   principal
** secondary

3. Status

Work in progress.

COMMISSION OF THE EUROPEAN COMMUNITIES ENERGY R & D PROGRAMME Objective: Solar Energy	Project or Sector: Photovoltaic conversion

Title:

Study of the morphology, fabrication and methods of increasing the efficiency of Gallium Arsenide Schottky-barrier solar cells.

Duration: 12 months Period: 1/12/76 - 1/12/77	Contract No: 163-76 ESEIR Project No: C/A/036(E)

Contractor: University College Cork

Address: Cork, Ireland

Head of Project: Dr. G. T. Wrixon

Description of research work

I. Objectives (aims)

The aim of the proposed research is to develop low cost high efficiency metal-GaAs solar cells. The Schottky approach has been chosen because of the construction simplicity and economy it offers. GaAs has been chosen as it is theoretically more efficient than Si and because its high surface recombination rate means that in a Schottky-barrier, where the junction is at the surface, e-h pairs formed near the surface are already in the junction field and can be split up before recombination. Because of the current cost of single crystal GaAs, such cells would have to work in a concentrator and thus the high temperature stability and performance of the cells are important.

2. Work Programme

(a) Investigation of various methods of anodizing Gallium Arsenide with the aim of perfecting a repeatable and reliable method of growing a uniform continuous native oxide on the semiconductor;

(b) Formation of MOS devices with oxide thicknesses of between 10 and 100Å with the aim of correlating η (as defined in the diode equation $I = I_o (\exp (qV/\eta\ell T)-1)$ with oxide thickness.

(c) Investigation of methods of laying down thin metal films using sputtering and evaporation techniques with the aim of obtaining a repeatable process for the production of uniform transparent films;

../..

Total cost:
  £   21,970
  u.a. 52,728
	E.C. Contribution: 57%
	  £ 12,500
	  u.a. 30,000

COMMISSION OF THE EUROPEAN COMMUNITIES ENERGY R & D PROGRAMME	Project or Sector:
Objective: Solar Energy	Photovoltaic conversion

2. Work Programme cont/...

(d) Design and practical realization of photomasks for current collecting fingers on cells;

(e) Formation of MOS solar cells and investigation of relative efficiency versus η measurement leading to value of optimum oxide layer thickness;

(f) Investigation of various anti-reflection coating - metal combinations and isolation of optimum configurations;

(g) Absolute efficiency measurements of optimum cell;

(h) In parallel with all of the above, a theoretical and experimental investigation of the effect of varying doping densities will proceed.

3. Status

(a) is almost complete;

(c) has been started;

(d) has been completed.

COMMISSION OF THE EUROPEAN COMMUNITIES ENERGY R & D PROGRAMME Objective: Solar Energy	Project or Sector: Photovoltaic conversion

Title: Process of deposition of single crystal silicon directly from vapour phase.	

Duration: 18 months Period: 1/9/76 - 22/3/78	Contract No: 186-77 ESI Project No: C/A/078(I)

Contractor: Montedison S.p.A. Address: Istituto Donegani, Via Fauser, Novara, Italy Head of Project: S. Pizzini	

Description of research work

I.   Objectives (aims)

The aim of the research is to show the feasibility of the CVD technique for depositing directly from a vapour phase, constituted by volatile silicon halogenides (SiH_4, $SiHCl_3$, SiH_2Cl_2), silicon single crystal rods of reasonable diameter.

It should also be demonstrated that the overall costs for the production of these single crystal bars should be at least a factor 2 less than the costs of CZ grown silicon single crystals.

2.   Work Programme

The research programme is divided into 3 stages:

(a)   Set-up of the deposition system, preliminary deposition runs using $SiHCl_3$ as the silicon halide.

(b)   Systematic deposition runs, in order to find the best deposition conditions, by varying the substrate temperature, the $\frac{SiHCl_3}{H_2}$ ratios (where H_2 is the carrier gas) and total flux.

../..

Total cost: Lit. 225,000,000 u.a. 360,000	E.C. Contribution: 31% Lit. 68,750,000 u.a. 110,000

COMMISSION OF THE EUROPEAN COMMUNITIES ENERGY R & D PROGRAMME Objective: Solar Energy	Project or Sector: Photovoltaic conversion

2. Work Programme cont/..

(c) Deposition runs with doping gases and/or with various halides
as the source of silicon.   Economic estimates of the process
will be performed in this period.

3. Status

Phase (a) has just ended with the set-up of the system (only the
automatic control of the temperature has to be improved) and several
deposition runs completed.   The influence of the $SiHCl_3/H_2$ ratio
and of the HCl etching has been investigated. Some partially single
crystal rods and two entirely single crystal rods (ϕ_{max} = 1 cm) have
been grown so far.

COMMISSION OF THE EUROPEAN COMMUNITIES ENERGY R & D PROGRAMME Objective: Solar Energy	Project or Sector: Photovoltaic conversion

Title:

Development of a Cadmiumselenide-Thin Film Solar Cell.

Duration: 18 months Period: 1/2/77 - 31/7/78	Contract No: 189-77 ESD Project No: C/A/085(D)

Contractor: Battelle Institut e.V.

Address: Postfach 900160, 6000 Frankfurt 90, Germany

Head of Project: Dr. Dieter Bonnet

Description of research work

   I. Objectives (aims)

   Development of a new thin film solar cell for terrestrial application
   consisting of a large-area contact between a CdSe layer and a
   transparent conducting electrode.

   The aim of this phase is to show that a sufficiently efficient CdSe
   solar cell can be made on thin film structures.   Consequently,
   a production technique suitable for a large area system shall be
   developed.

   2. Work Programme

   (a) Production of thin CdSe films, optimization of doping and doping
        profile;

   (b) Optimization of transparent cover electrodes and deposition onto
        CdSe-films;

   (c) Development and deposition of current collection contacts;

   (d) Investigations into chemical and crystalline structures of the
        films.

   3. Status

   Work in progress.

Total cost:
     DM.   454,200
     u.a.   124,098
	E.C. Contribution: 50%
	     DM. 227,100
	     u.a. 62,049

COMMISSION OF THE EUROPEAN COMMUNITIES ENERGY R & D PROGRAMME Objective: Solar Energy	Project or Sector: Photovoltaic conversion

Title:	Studies of the physical principles limiting the performance of cadmium sulphide and other thin film solar cells and of the most suitable methods of fabricating these cells to realise enhanced efficiencies and reliability and significantly reduce the cost of production.

Duration: 18 months Period: 1/10/76 - 31/3/78	Contract No: 191-76 ESUK Project No: C/A/122(UK)

Contractor: International Research & Development Co. Ltd.

Address: Fossway Newcastle-upon-Tyne, NE6 2YD, United Kingdom

Head of Project: Dr. L. Clark

Description of research work

I. Objectives (aims)

(a) Better qualification of the 'Clevite' process, with the aim of improving efficiency, reproducibility and reliability of cells, by investigating the effects of impurities in the starting materials and by controlling the stoichiometry of the Cu_2S, as well as by gaining more basic understanding of the role of surface states in limiting cell efficiencies;

(b) Evaluation of alternative processes for the production of the cadmium sulphide layers and for barrier formation which, if successful, could offer more efficient use of cadmium, improved efficiencies and stability, and simplification of the production process. The main effort will be on anodic sulphurization of cadmium.

2. Work Programme

The construction of necessary apparatus for the work outlined in 'objectives' above, followed by the deposition of and evaluation of thin films of CdS, and the fabrication and evaluation of complete solar cells. In particular, films and cells will be assessed by the techniques outlined in the viewgraphs of the first oral presentation (March 1st). Various appropriate starting materials will be evaluated and used.

../..

Total cost: £ 55,606 u.a. 133,454	E.C. Contribution: 50% £ 27,803 u.a. 66,727

COMMISSION OF THE
EUROPEAN COMMUNITIES

ENERGY R & D PROGRAMME

Objective: Solar Energy

Project or Sector:

Photovoltaic conversion

3. Status

Apparatus has been constructed for CdS film assessment (resistivity Hall effect, in dark and illuminated; microscopic examination) and for Cu_2S and cell assessment (resistivity; short circuit current versus temperature in various ambients).

CdS layers have been produced, by vacuum deposition, on to molybdenum, glass and metallised plastic substrates and are being used to evaluate the film characterisation apparatus.

A work programme has been drawn up for the anodic sulphurisation studies, but awaits the completion of a new fume cupboard (due March 18th) before practical work with the toxic solutions can begin.

COMMISSION OF THE EUROPEAN COMMUNITIES ENERGY R & D PROGRAMME Objective: Solar Energy	Project or Sector: Photovoltaic conversion

Title:

Study and production of photovoltaic cells made of heterostructures III-V compounds deposited in thin films from organo-metallic and coordination compounds on alumina thin foil.

Duration: 18 months Period: 12/76 - 6/78	Contract No: 193-77 ESF Project No: C/A/029a(F)

Contractor: Centre National d'Etudes Spatiales (C.N.E.S.)

Address: 18, Avenue Edouard Belin, 31055 Toulouse Cedex, France

Head of Project: J. Cacheax (Departement Electronique Physique)

Description of research work

I. Objectives (aims)

Solar cells in III-V multiple semiconductor compounds

- physical optimization of structures;
- economic optimization of manufacturing processes.

2. Work Programme

A. Orientation report: this report will contain:

- a description of the crystal growing apparatus (heteroepitaxy);
- preparation of substrates;
- characterization methods;
- synthesis and preparation of organo-metallic compounds;
- a presentation of the structures studied (first-generation structure with forbidden band profile and flat band transition - Ga As (Ga Al As).

../..

Total cost:
   FF. 1,100,000
   u.a. 200,000
	E.C. Contribution: 30%
	   FF. 333,300
	   u.a. 60,000

COMMISSION OF THE EUROPEAN COMMUNITIES ENERGY R & D PROGRAMME Objective: Solar Energy	Project or Sector: Photovoltaic conversion

2. Work Programme cont/..

B. Interim study report:

- design and optimization of the first-generation cell;

- study of the synthesis of coordination material;

- trial under concentration conditions.

C. Interim study report:

- design and optimization of the second generation cell;

- results of the coordination product test;

- concentration study.

D. Economic study report.

E. Final study report.

3. Status

- Study of crystal formation by CVD- OM on alumina substrate;

- Model made of the first generation structure;

- Study of the concentrators;

- Characterization apparatus set up.

COMMISSION OF THE EUROPEAN COMMUNITIES ENERGY R & D PROGRAMME Objective: Solar Energy	Project or Sector: Photovoltaic conversion

Title:

Study of manufacturing processes for pure organo-metallic substances used in the fabrication of semicondictor films by chemical deposition from the vapour phase.

Duration: 18 months Period: 12/76 - 6/78	Contract No: 193-77 ESF Project No:

Contractor: Institut National de Recherche Chimique Appliquée (I.R.C.H.A.)

Address: 16, rue Jules César, 75012 Paris, France

Head of Project: F. Mathey

Description of research work

I. Objectives (aims)

- Study and manufacture of organo-metallic materials necessary for the crystal growing of III-V compounds;

- Study of purification;

- Study of physical properties.

2. Work Programme

First stage:

Preparation of basic organo-metallic substances;

Determination of purity;

Selection of preparation methods for the optimium elimination of impurities deleterious to electronic use.

Second stage:

(a) Determination of purification methods;

(b) Study of association compounds, either in order to effect the chemical decomposition directly in the vapour phase or as an intermediary to purify the products.

../..

Total cost: FF. 370,000 u.a. 67,000	E.C. Contribution: 30% FF. 111,100 u.a. 20,000

COMMISSION OF THE EUROPEAN COMMUNITIES ENERGY R & D PROGRAMME	Project or Sector:
Objective: Solar Energy	Photovoltaic conversion

2. Work Programme cont/..

Third stage:

Changeover to small-scale pilot project.

3. Status

Production of trimethyl-gallium completed.

COMMISSION OF THE EUROPEAN COMMUNITIES ENERGY R & D PROGRAMME Objective: Solar Energy	Project or Sector: Photovoltaic conversion

Title:

GaAs-(GaAl)As solar cells to be used under concentrated solar light conditions.

Duration: 18 months Period: 1/7/76 - 31/12/77	Contract No: 194-76 ESI Project No: C/A/075(I)

Contractor: Cise SpA

Address: C.P. 3986 - 20100 Milano, Italy

Head of Project: Prof. Enrico Cerrari

Description of research work

I. Objectives (aims)

The research should allow a better definition of the role of GaAs-(GaAl)As heterostructures as a material for a number of solar cell terrestrial applications. The research aims at the construction of the prototype; the device is, however, made in view of a subsequent industrial engineering of the production method.

The ultimate purpose of the research is, of course, to get competitive costs for the energy thus obtained, at least in some terrestrial applications.

2. Work Programme

The research programme, consisting of two successive phases, can be summarized as follows:

Phase 1 - In cooperation with SELENIA, a preliminary project of the cell will be made in the aim at obtaining the best evaluation of the material critical parameters. The doping levels and the junction depth, suitable to give the maximum conversion efficiency, will be especially investigated.

The possibility of modification of an apparatus for (GaAl)As liquid epitaxy, already at hand in our laboratory, will be experimentally tested in order to establish the features of a subsequent "ad hoc" reactor.

../..

Total cost:
    Lit. 326,700,000
    u.a.   522,720
	E.C. Contribution: 7%
	    Lit. 21,875,000
	    u.a. 35,000

COMMISSION OF THE EUROPEAN COMMUNITIES ENERGY R & D PROGRAMME Objective: Solar Energy	Project or Sector: Photovoltaic conversion

2. Work Programme cont/..

Phase 2 — The information obtained during Phase 1, will allow the development of a laboratory prototype of a first generation reactor for liquid epitaxy, fit for supplying suitable GaAs-(GaAl)As heterostructures in accordance with the needs displayed by the preliminary project of the cell.

An oriented evaluation of the obtained material will be made in order to get the information for the growth process optimization.

After consistency of the material critical parameters with the values from the project has been checked, the setting up of photovoltaic elements will take place as a last operation test.   There will be, then, the setting up of the reactor in its final stage, which will supply the material necessary for the development of the cell prototype.   The research on the device will be carried on at the same time.

The ultimate cell configuration will be designed on the basis of a study of the cell-concentrator system.   Due account will be given to the interdependent parameters, that is: concentration ratio, working temperature and conversion efficiency.

In cooperation with MONTEDISON, Istituto Donegani, various types of surface passivants and antireflecting layers will be tested.

Passivants and antireflecting layers, will also be investigated with SELENIA which will also cooperate in the optimization of the cell upper contact.

3. Status

Work in progress.

COMMISSION OF THE EUROPEAN COMMUNITIES ENERGY R & D PROGRAMME Objective: Solar Energy	Project or Sector: Photovoltaic conversion

Title:

Investigation of Ion Implantation as a technique
suitable to fabricate high efficiency silicon solar
cells.

Duration: 18 months Period: 1/7/1976 - 31/12/1977	Contract No: 195-76 ESI Project No: C/A/081(I)

Contractor: Laboratorio di Chimica e Technologia dei Materiali e dei
Componenti per l'Elettronica (LAMEL) del C.N.R.

Address: Via Castagnoli 1, 40126 Bologna, Italy

Head of Project: Dr. Giovanni Soncini

Description of research work

I. Objectives (aims)

Study of the possible advantages offered by Ion Implantation to
produce high efficiency silicon solar cells.   Due to the high dosage
required ($>10^{15}$ ions/cm^2), particular attention will be paid to the
radiation damage and to its annealing characteristics, as related to
the parameters of the process (i.e., crystal orientation, random or
channeling implantation etc.).

Prototypes of implanted cells will be fabricated and characterized by
measuring their spectral response.   The experimental results will be
interpreted by using a unidimensional computer programme to simulate
the solar cell behaviour.

2. Work Programme

(a) Fabrication of silicon solar cell structures by ion implantation
through thermal oxide layers.   Development of a theoretical
model to compute implanted doping profiles and its experimental
verification;

(b) Set-up of measurement techniques to evaluate the spectral response
of the cell within the wavelength range of interest for the
photovoltaic energy conversion.   Development of a numerical
programme to analyse the response curve;

(c) Comparative evaluation of ion implantation and thermal diffusion
as techniques suitable for large scale production of high efficiency
silicon solar cells.   ../..

Total cost:
    Lit.   163,700,000
    u.a.   261,920
	E.C. Contribution: 23%
	    Lit. 37,500,000
	    u.a. 60,000

COMMISSION OF THE EUROPEAN COMMUNITIES	Project or Sector:
ENERGY R & D PROGRAMME	Photovoltaic conversion
Objective: Solar Energy	

3. Status

The research activity, both theoretical and experimental, is going on as foreseen.

COMMISSION OF THE EUROPEAN COMMUNITIES	Project or Sector:
ENERGY R & D PROGRAMME	Photovoltaic conversion
Objective: Solar Energy	

Title:

Feasibility study on the use of CdTe for solar cell fabrication.

Duration: 18 months	Contract No: 206-76 ESF
Period: 1/7/76 - 31/12/77	Project No:

Contractor:
1. Centre National de la Recherche Scientifique (CNRS).
2. Office National d'Etudes et de Recherches Aérospatiales
3. Université Paul Sabatier. (ONERA).

Address:
1. Paris; 2./3. Toulouse : France

Head of Project:

Description of research work

I. Objectives (aims)

Research of the elaboration conditions of CdTe thin films and realization of an internal potential barrier which might lead to improved photocells.

2. Work Programme

There is a three part programme:

(a) technological realization of materials and structures;

(b) physical analysis and description of the electrical properties of the structures;

(c) description and analysis of the structures when used as photocells.

3. Status

Work in progress.

Total cost: FF. 1,014,100 u.a. 182,639	E.C. Contribution: 16% FF. 161,100 u.a. 29,014

Project D

Photochemical, photoelectrochemical
and photobiological processes

COMMISSION OF THE EUROPEAN COMMUNITIES ENERGY R & D PROGRAMME Objective: Solar Energy	Project or Sector: Photochemical, photoelectrochemical and photobiological processes

Title: Kinetic factors controling the efficiency of photosystem II.	

Duration: 12 months Period: 1/12/76 - 30/11/77	Contract No: 013-76 ESF Project No:

Contractor: CNRS - Laboratoire de Photosynthèse

Address: 15, quai Anatole France, 75700 Paris

Head of Project: J. Lavorel, Directeur de Recherche CNRS,
Directeur du Laboratoire de Photosynthèse du CNRS

Description of research work

I. Objectives (aims)

The functioning of system II of photosynthesis in higher plants as witnessed by the O_2 emission in sequence of flashes is currently understood (Kok's model) as implying a cyclic mechanism of positive charges accumulation, subject to two effects: "misses" (deficit of quantum yield) and "double hits". The miss factor is important since it would limit, if the interpretation is right, the overall efficiency of the photosynthetic apparatus.

Our objective is to verify this interpretation. A working hypothesis is that misses in general are due to "dumping" factors which do not necessarily lead to a deficit in quantum yield.

2. Work Programme

- Experimental factors modifying the yield and the dumping coefficient of O_2 emission in sequence of flashes. Analysis of the result with the matrix method;

- Investigation of models involving different causes of dumping;

- Correlation between misses, deactivation, initial distribution of S states;

- Double hits in rectangular laser flashes;

- Dark triggered luminescence.

../..

Total cost: FF. 204,720 u.a. 36,860	E.C. Contribution: 50% FF. 102,360 u.a. 18,430

COMMISSION OF THE
EUROPEAN COMMUNITIES

ENERGY R & D PROGRAMME

Objective: Solar Energy

Project or Sector:

Photochemical, photoelectrochemical
and photobiological processes

3. Status

We have already shown that KOK's interpretation of the dumping
factor as "misses" is probably wrong (for instance, cases exist
where both the dumping and yield decrease, which is contradictory
to the above interpretation).   Several models may explain these
results, one in particular implying a mobility of the O_2 evolving
system with respect to the System II centre.

COMMISSION OF THE EUROPEAN COMMUNITIES ENERGY R & D PROGRAMME Objective:　　Solar Energy	Project or Sector: Photochemical, photoelectrochemical and photobiological processes

Title: Study of primary reactions of photosynthesis in green plants by means of rapid flash absorption spectroscopy.	

Duration:　　12 months Period:　29/9/76 - 28/9/77	Contract No:　　014-76　ESF Project No:

Contractor: Commissariat à l'Energie Atomique (C.E.A.)

Address: Dept. de Biologie du CEN-SACLAY
　　　　　29 rue de la Federation, Paris, France

Head of Project:　　Paul Mathis

Description of research work

I.　Objectives (aims)

The aim of our work is to elucidate the mechanism by which the energy of light is converted into chemical free energy at special locii ("reaction centres") in green plants.　Two types of reaction centres (Photosystem-1 and Photosystem-2) have been evidenced.　For each of them we wish to contribute:

(i)　to the chemical identification of the reaction partners (primary and secondary electron donors and acceptors) of the rate and path of electron transfer reactions;

(ii) to the identification of the excited states which are involved in the primary photochemical processes.

2.　Work Programme

The preceding objectives will be achieved by the use of fast absorption spectroscopy, which allows the detection of the effects of a short actinic flash thanks to the changes in the absorption spectrum of involved species.　We propose to focus our effort on these aspects:

(i)　a study of the reactions on the donor side of Photosystem-2;

(ii)　the behaviour of Photosystem-1 at low redox potentials, at which the electron path is selectively restricted to a limited number of electron acceptors;

../..

Total cost: 　FF.　204,720 　u.a.　36,860	E.C. Contribution:　　50% 　FF.　102,360 　u.a.　18,430

COMMISSION OF THE EUROPEAN COMMUNITIES ENERGY R & D PROGRAMME Objective: Solar Energy	Project or Sector: Photochemical, photoelectrochemical and photobiological processes

2. Work Programme cont/..

(iii) the properties of excited triplet states (of chlorophyll and carotenoid) in the photosynthetic membrane.

3. Status

Promising (although still preliminary) results have been obtained, especially concerning Photosystem-1.

In Photosystem-1 particles our results allow us to propose that two electron acceptors are more primary than P-430.   Their back-reaction with $P+_{700}$ occurs in times of (respectively) 300 μs and 5 μs.

COMMISSION OF THE EUROPEAN COMMUNITIES ENERGY R & D PROGRAMME Objective: Solar Energy	Project or Sector: Photochemical, photoelectrochemical and photobiological processes

Title: Study of the electron transport in the photosynthesis.

Duration: 12 months Period: 13/9/76 - 12/9/77	Contract No: 015-76 ESF Project No:

Contractor: Institut de Biologie Physico-chimique, Fondation Edmond de Rothschild

Address: 13, rue Pierre et Marie Curie - 75005 Paris, France

Head of Project: P. Joliot

Description of research work

I. Objectives (aims)

Characterization of the components of the electron transport chain. Photosystem II: The number and function of the redox components associated with the donor and acceptor sides of PSII has recently undergone a re-evaluation in our and other laboratories. At least three donors are directly connected to the photoactive chlorophyll and recent results, obtained in our laboratory, established the existence of two primary acceptors. The function of the different donors is currently being studied by fluorescence and oxygen detecting techniques, the localisation within the photosynthetic membranes of these various components will be studied by detection of the transmembrane electric field.

PhotosystemI: A study is in progress concerning the electron transfer reactions occuring on the donor and acceptor sides of PS I. The experimental technique used is that of a rapid flash detection spectrophotometer which permit measurement of the redox state of the various components and of the field indicating absorbance change.

A final programme of research concerns the light harvesting antenna. These studies are being carried out on Chlamydomonas mutants and on Cyanidium. A study is also in progress comparing functional measurements (fluorescence, O_2 evolution, action spectra) to structure as determining by freeze fracture electron microscopy. Research on the mode of action of DCMU is also projected for the near future. A programme for the selection of mutants blocked on the donor side of Photosystem II is being persued.

Total cost:
  FF. 204,720
  u.a. 36,860
	E.C. Contribution: 50%
	  FF. 102,360
	  u.a. 18,430

COMMISSION OF THE EUROPEAN COMMUNITIES ENERGY R & D PROGRAMME Objective: Solar Energy	Project or Sector: Photochemical, photoelectrochemical and photobiological processes

2. Work Programme

Photosystem II: Fluorescence studies have permitted the demonstration that two electron acceptors are connected in parallel to the reaction centres.   Both acceptors are quenchers of fluorescence in their oxidized state and probably only one is connected to the plasto-quinone pool.

Spectrophotometric detection of the field detecting absorbance changes suggest that the photoactive chlorophyll is located close to the outside of the thylakoid membrane.   Oxygen and fluorescence experiments performed under anaerobic atmospheres have permitted a complete characterization of the quilibrium constants relating the primary and secondary acceptors.

Photosystem I: The sequence of electron transfer occuring between the plastoquineone pool and reaction centre chlorophyll P700 have been elecidated by spectrophotometric measurements.   Photosystem I also participates in an electrogenic loop other than linear and cyclic electron transport.

Antenna: Fluorescence and oxygen measurements performed on Cyanidium have shown that only a fraction of the Photosystem II centres are directly connected to the phycobilisomes.   Investigation of a Chlamydomonas mutant, lacking in pigment complex I, demonstrates an uphill energy transfer from System I to System II.

3. Status

Work in progress.

COMMISSION OF THE EUROPEAN COMMUNITIES ENERGY R & D PROGRAMME Objective: Solar Energy	Project or Sector: Photochemical, photoelectrochemical and photobiological processes

Title: Fundamental studies of photochemical and photobiological systems for the use of solar energy.	

Duration: 12 months Period: 13/9/76 - 12/9/77	Contract No: 016-76 ESUK Project No:

Contractor: Imperial College of Science and Technology Address: Dept. of Botany, Imperial College of Science and Technology, London S.W.7. 2AZ. Head of Project: Dr. J. Barber	

Description of research work

I. Objectives (aims)

To investigate the organisation of the photosynthetic light harvesting pigments and elucidate the mechanisms of energy transfer between them. To gain a better understanding of the exciton migration between the photosynthetic units of oxygen evolving organisms and in particular study the control mechanism which functions to maximise photosynthetic efficiency by optimising the quantal distribution to the two photosystems. To relate the molecular mechanism for regulating this quantal distribution with the normal physiological functioning of the intact chloroplast with particular emphasis on the part played by light induced changes in the cationic composition of the chloroplast.

2. Work Programme

(i)   To study energy transfer mechanisms in photosynthetic material by the use of ultra-short pulsed spectroscopy;

(ii)   To investigate the role of metal cations and proton gradients in bringing about changes in quantal distribution between photosystem one and two by means of studying chlorophyll fluorescence yield and lifetime changes;

(iii) To gain an understanding of the physical mechanisms controlling the changes in quantal distribution and chlorophyll fluorescence;

(iv)   To carry out detailed analyses of the ionic composition of chloroplasts under a variety of conditions and relate these   ../..

Total cost: £ 23,050 u.a. 55,320	E.C. Contribution: 50% £ 11,525 u.a. 27,660

COMMISSION OF THE
EUROPEAN COMMUNITIES

ENERGY R & D PROGRAMME

Objective: Solar Energy

Project or Sector:

Photochemical, photoelectrochemical
and photobiological processes

2. Work Programme cont/...

(iv) with the above studies.

3. Status

(a) Picosecond fluorescence lifetime and yield measurements have been
made with sub-chloroplast particles enriched in photosystem one
and two activity as well as with light harvesting complexes devoid
of reaction centres and some of the work has been published (1).
Detailed studies have also been made into the mechanism of energy
transfer between the light harvesting pigments of the red algae
Porphyridium (2,3).   Chlorophyll lifetime measurements are being
carried out with intact organisms driven into either State one or
State two by preirradiation.

(b) A thorough study has been made of selectivity, competition and
antagonism of a range of metal cations at inducing changes in
chlorophyll fluorescence yield of isolated chloroplasts (4).
Picosecond fluorescence yield conditions in order to elucidate
the interaction of the various chlorophylls in controlling the mode
and direction of energy transfer between the two photosystems.

(c) The above studies have given rise to the concept that the control
of quantal distribution between the two photosystems is regulated
by changes in the diffuse electrical layer adjacent to the
negatively charged thylakoid membrane.   In particular it seems
that the membrane conformational changes which give rise to the
fluorescence effects result from changes in the space charge
density within a few ångstroms of the membrane surface.   The
concepts have been published (4,9) and have led to a study of
fluorescence changes of 9-aminoacridine which has served as a
cationic probe of the double layer (6).

(d) A thorough analyses of ionic distribution in isolated chloroplast
is underway by using the multielemental technique of neutron
activation.   Particle electrophoresis investigations have been
initiated in order to gain knowledge about the density and
properties of chloroplast membrane surface charges.   This work has
greatly benefited by the discovery of a new technique of obtaining
isolated chloroplast suspensions with a high degree of intactness
(i.e., with their outer membranes) ref.7.

(e) Experiments are being designed to understand the physiological
control mechanism of quantal distribution particularly in relation
to the generation of light induced pH and electrical gradients
between the stromal and intrathylakoid compartments of the
chloroplast (7).

COMMISSION OF THE EUROPEAN COMMUNITIES ENERGY R & D PROGRAMME Objective: Solar Energy	Project or Sector: Photochemical, photoelectrochemical and photobiological processes

Title: (i) Elucidation of the primary processes of photosyntheses;
(ii) The construction of model systems in vitro which operate in a similar manner and which have promise as economic methods for the collection and utilisation of solar energy.

Duration: 12 months Period: 1/10/76 - 30/9/77	Contract No: 017-76 ESUK Project No:

Contractor: Members of the Royal Institution of Great-Britain

Address: 21, Albemarle Street,
London W1X 4BS, United Kingdom

Head of Project: Professor Sir George Porter, F.R.S.

Description of research work

I. Objectives (aims)

The objectives of the research are to imitate the photosynthetic process in vitro so as to store solar energy as a chemical fuel. Particular attention is being paid to the photodissociation of water by visible light using mechanisms based on photosystem II of plant photosynthesis.

2. Work Programme

The work of the group includes (a) in vivo studies of photosynthesis using flash photolysis and fluorescence methods over the time range from microseconds to picoseconds, (b) the preparation of model systems which incorporate chlorophyll and redox systems in lipid membranes and vesicles, and (c) photochemical studies of electron transfer reactions with particular attention to those incorporating complexes of manganese. Most of the effort of those employed on the project has been in this last area.

3. Status

The results obtained with manganese complexes, particularly those of phthalocyanine are promising. Phthalocyanine is a very stable molecule and is highly resistant towards oxidation. Manganese forms stable phthalocyanine complexes with +2, +3 and +4 formula oxidation states, although in aerated solution the +3 state normally predominates, and all the complexes have intense absorption bands in the 600-750 nm region. In the presence of a quinone as electron acceptor, $Mn^{II}Pc$ undergoes facile photooxidation to give the Mn^{III} complex and reduced quinone:

../..

Total cost:
    £   23,050
   u.a. 55,320
	E.C. Contribution: 50%
	    £ 11,525
	  u.a. 27,660

COMMISSION OF THE EUROPEAN COMMUNITIES ENERGY R & D PROGRAMME Objective: Solar Energy	Project or Sector: Photochemical, photoelectrochemical and photobiological processes

3. Status cont/..

$$*Mn^{II}Pc \; + \; Q \; \xrightarrow{H^+} \; Mn^{III}Pc + QH^\cdot$$

Further irradiation results in oxidation of Mn^{III} to Mn^{IV}.

$$*Mn^{III}Pc \; + \; Q \; \xrightarrow{H^+} \; Mn^{IV}Pc \; + \; QH^\cdot$$

There is some evidence that Mn^{IV} complexes preferentially adopt a binuclear structure using oxygen functions as bridging ligands and this is probably the case with phthalocyanine complexes.

$$2Mn^{IV}Pc \rightleftharpoons \; Mn^{IV}Pc \longrightarrow \; Mn^{IV}Pc$$

Thus, irradiation results in oxidation of Mn^{II} to Mn^{IV} which, in the form of a binuclear complex, can act as a four electron acceptor and so accommodate the full electron balance necessary to produce one molecule of oxygen from water.

The binuclear Mn^{IV} complex is a powerful oxidising species and can be converted readily into the original $Mn^{II}Pc$. A particularly efficient reductant for this reaction is durohydroquinone and it is essential that the reduced form of the electron acceptor does not act as a reductant for Mn^{IV}. When this factor is eliminated, the $Mn^{IV}Pc$ complex is available for reaction with the aqueous substrate. Under these conditions, reduction to $Mn^{II}Pc$ takes place although the rate of reduction is very much dependent upon the pH of the solution. At high pH (11), Mn^{IV} is stable with respect to both water and hydroxide ions. At intermediate pH values, some reduction does occur but it is a slow process whilst in acidic solution, reduction to Mn^{II} is very efficient and proceeds via oxygen evolution.

$$2H_2O + \left[Mn^{IV}Pc \text{---} Mn^{IV}Pc \right] \longrightarrow O_2 + 4H^+ + 2Mn^{II}Pc$$

Protons formed during the redox reactions are collected by the reduced electron acceptor so that the quinone is converted into semiquinone.

$$H^+ + Q + e \longrightarrow QH^\cdot$$

COMMISSION OF THE EUROPEAN COMMUNITIES ENERGY R & D PROGRAMME Objective: Solar Energy	Project or Sector: Photochemical, photoelectrochemical and photobiological processes

3. Status cont/..

The semiquinone radical has been observed by flash spectroscopy for the reactions between duroquinone and both $Mn^{II}Pc$ and $Mn^{III}Pc$. It decays by a second order process which is most probably disproportionation yielding hydroquinone.

$$2QH^{\bullet} \longrightarrow Q + QH_2$$

Ultimately, the hydroquinone should be converted into the original quinone and hydrogen gas although we have not yet attempted this process.

COMMISSION OF THE EUROPEAN COMMUNITIES ENERGY R & D PROGRAMME Objective: Solar Energy	Project or Sector: Photochemical, photoelectrochemical and phtoobiological processes

Title:

Photoenergy conversion in artifical systems.

Duration: 12 months Period: 13/9/76 - 12/9/77	Contract No: 018-76 ESUK Project No:

Contractor: University of Bristol

Address: Senate House, Bristol BS8 1TH, United Kingdom

Head of Project: Dr. A.R. Crofts

Description of research work

I. Objectives (aims)

To study the photoenergy conversion in artificial systems.

2. Work Programme

(a) Reconstitution of the proton pumping activity of photochemical reaction centres by incorporating the purified proteins in artificial lipid vesicles;

(b) Incorporation of reaction centres in an organic phase (hexane);

(c) Studies of the physical structure of reaction centres in hexane using X-ray and neutron low angle scattering techniques;

(d) Preparation of bi-layers from monolayers using the hexane dissolved reaction centres, with a view to studying the photoelectric properties of the reconstituted reaction centre;

(e) Preparation of multi-layered stacks from monolayers with a view to (i) studying the structure of the reaction centre and orientation of chromophores, and (ii) forming microphotobatteries of cells using the photoelectrogenic properties of oriented reaction centres.

../..

Total cost:
  £   23,050
  u.a. 55,320
	E.C. Contribution: 50%
	  £ 11,525
	  u.a. 27,660

COMMISSION OF THE EUROPEAN COMMUNITIES ENERGY R & D PROGRAMME Objective: Solar Energy	Project or Sector: Photochemical, photoelectrochemical and photobiological processes

3. Status

The work under headings (a) – (c) above is already well advanced, and we will continue to pursue these objectives along the lines of our current programme.   Projects (d) and (e) are at present held up by the development of suitable apparatus.   We are able to form stable monolayers containing reaction centres, but are not yet able to characterise these with respect to pressure/area isotherms, etc.   We are characterising the biochemical properties of the hexane solution so as to be able to optimise protein to lipid ratios, and thus avoid excess lipid in the monolayers.

COMMISSION OF THE EUROPEAN COMMUNITIES ENERGY R & D PROGRAMME Objective: Solar Energy	Project or Sector: Photochemical, photoelectrochemical and photobiological processes

Title:

Investigation of the properties of the primary electron acceptor complex of photosystem I.

Duration: 12 months Period: 22/7/76 - 21/7/77	Contract No: 019—76 ESUK Project No:

Contractor: University College, London

Address: Dept. of Botany and Microbiology, University College, Gower St., London WCiE 6BT, United Kingdom

Head of Project: Dr. C.W. Evans

Description of research work

I. Objectives (aims)

The conversion of light energy into chemical energy in the reaction centre of photosystem I involves the photooxidation of a reaction centre chlorophyll (P700) and the reduction of an electron acceptor complex containing two bound iron—sulphur centres and a chemically unidentified component "X" which is the primary electron acceptor. The research programme is designed to define (a) the chemical identity of "X" and other components of the reaction centre, (b) the oxidation—reduction characteristics of these components and (c) the structure of the reaction centre and the minimal complex required for energy trapping.

2. Work Programme

The properties of the photosystem I reaction centre are being investigated using low temperature e.p.r. spectrometry and absorption spectrophotometry. Quantitative determination of the components of the reaction centre and of their oxidation reduction potential are being made using these techniques. Most of the work on PSI has been done using partially purified membrane fractions. Techniques will be developed for the isolation of defined reaction centre preparations. The chemical composition of the primary electron acceptor is being investigated using a number of different techniques.

3. Status

We have developed techniques for the large scale preparation of PSI reaction centres free of electron transport components other than

../..

Total cost: £ 23,050 u.a. 55,320	E.C. Contribution: 50% £ 11,520 u.a. 27,660

COMMISSION OF THE EUROPEAN COMMUNITIES ENERGY R & D PROGRAMME Objective: Solar Energy	Project or Sector: Photochemical, photoelectrochemical and photobiological processes

3. Status cont/..

those of the reaction centre.   The P700: chlorophyll ratio is however 1:30, work is continuing in an attempt to obtain preparation with very much less light harvesting chlorophyll.

Quantitative e.p.r. measurements have been used to show that P700, X and iron-sulphur centres A & B are present in the reaction centre in equivalent amounts.   The redox potential of P700 has been determined as $E = +375mV$.   Centre A has $E_m = -550mV$ and Centre B has $E = -590m V$.   The potential of "X" is too low to measure by standard techniques and new methods must be developed.

COMMISSION OF THE EUROPEAN COMMUNITIES ENERGY R & D PROGRAMME Objective: Solar Energy	Project or Sector: Photochemical, photoelectrochemical and photobiological processes

| Title:

Biophotolysis of water for hydrogen production via natural and artificial catalytic systems. ||

Duration: 12 months Period: 17/6/77 - 16/6/77	Contract No: 020-76 ESUK Project No:

Contractor: University of London King's College

Address: Dept. Plant Sciences, 68 Half Moon Lane,
              London SE24 9JF

Head of Project: Prof. D.O. Hall

Description of research work

I. Objectives (aims)

Certain algae have the capacity of producing H_2 gas in the light using H_2O as the source of electrons and the enzyme hydrogenase to evolve the H_2. The aim of this project is to study the inherent properties of an in vitro system to see if it could ultimately be developed into an energy conversion system utilising the sun as a primary source.

Four aspects of the system are being investigated:

(a) stabilization and storage of the chloroplast membranes;

(b) characterization and stabilization of hydrogenases from a range of organisms;

(c) replacement of various parts of the system with artificial analogues;

(d) construction of a two stage apparatus whereby O_2 is evolved from water in one compartment and H_2 in another.

2. Work Programme

The investigators are studying an in vitro system for biophotolysis of water. At its simplest this system contains chloroplasts, an electron mediator and hydrogenase in a buffered system. This system is being characterized with reference to rate, duration and efficiency of hydrogen production. It has been found that the two major problems are the lability of the chloroplasts and the oxygen sensitivity of the hydrogenase.

../..

Total cost:
    £   23,050
    u.a. 55,320
	E.C. Contribution: 50%
	    £ 11,525
	    u.a. 27,660

COMMISSION OF THE EUROPEAN COMMUNITIES ENERGY R & D PROGRAMME Objective: Solar Energy	Project or Sector: Photochemical, photoelectrochemical and photobiological processes

2. Work Programme cont/...

Many of the analogues that could replace parts of the system are stable only in organic solvents.   A study has been carried out on the effect of one of these solvents on the H_2 evolving system.

Various organisms have been tested for hydrogenase activity.

3. Status

The chloroplast hydrogenase system has been fairly well characterized. The investigators are now concentrating on:

(a)   a serious study of chloroplast stability;

(b)   a survey of a wide range of organisms for hydrogenase activity;

(c)   physico-chemical characterization of hydrogenases; (sections (b) and (c) are an attempt to find an oxygen-insensitive hydrogenase);

(d)   a study of the interaction of various manganes and iron complexes with the chloroplast and hydrogenase.

COMMISSION OF THE EUROPEAN COMMUNITIES ENERGY R & D PROGRAMME Objective: Solar Energy	Project or Sector: Photochemical, photoelectrochemical and photobiological processes

Title:

Light dependent hydrogen evolution by isolated chloroplasts and by algae.

Duration: 12 months Period: 13/9/76 - 12/9/77	Contract No: 021-76 ESD Project No:

Contractor: Ruhr-Universität Bochum

Address: Ruhr Universität Bochum, Lehrstuhl für Biochemie der Pflanzen, Postfach 10 21 48, 4630 Bochum 1, Germany

Head of Project: Prof. Dr. A. Trebst

Description of research work

I. Objectives (aims)

In photosynthetic electron flow water is photooxidize to yield oxygen and very electronegative reducing power.   It should be possible to couple the later onto a bacterial hydrogenase.   This way a photosynthetic system should split water to oxygen and hydrogen in the light with rates of the photosynthetic capacity of plants i.e., 200 μmoles hydrogen/mg chlorophyll and h at saturating light.

2. Work Programme

We have proposed to study the coupling of hydrogenase systems to the photosynthetic apparatus in isolated chloroplasts as well as in intact algae.   The first might have the advantage, that the various regulating steps in photosynthesis are overcome and therefore the photosynthetic rates are increased.   In the later, intact algae may be used for hydrogen evolution.

In the first half year of the programme we have concentrated on the chloroplast system with the aim to optimize hydrogen evolution by a chloroplast system from a plant plus a bacterial hydrogenase.

3. Status

We did succeed in the coupling of photosystem I of isolated chloroplasts to a Clostridium hydrogenase with a rate of 125 μmol hydrogen/mg chlorophyll/h, evolved in saturating light, i.e., rates approaching the maximal rate of the photosynthetic apparatus.   The conditions were
../..

Total cost:
   DM.   118,950
   u.a.   32,500
	E. C. Contribution: 50%
	   DM. 59,475
	   u.a. 16,250

COMMISSION OF THE EUROPEAN COMMUNITIES ENERGY R & D PROGRAMME Objective: Solar Energy	Project or Sector: Photochemical, photoelectrochemical and photobiological processes

methylviologen as mediator between photosystem I and the hydrogenase, N-tetramethyl-p-phenylenediamine as the electron donor and — particularly important — a closed chloroplast thylakoid vesicle.   The conditions indicate that back reactions of reduced methylviologen to the donors site of photosystem I and cyclic electron flow compete with the non-cyclic electron flow to hydrogen.   Hydrophilic, polar mediators and closed vesicles prevent such back reactions.   The results are discussed in detail and in relation to the literature in a paper by D. Hoffmann, R. Thauer and A. Trebst "Photosynthetic hydrogen evolution by spinach chloroplasts coupled to a Chlostridium hydrogenase" to appear in the next issue of Z. f. Naturforschung.

COMMISSION OF THE EUROPEAN COMMUNITIES ENERGY R & D PROGRAMME Objective: Solar Energy	Project or Sector: Photochemical, photoelectrochemical and photobiological processes

Title:

      Studies on the mechanism of a light driven proton pump: bacteriorhodopsin.

Duration: 12 months Period: 15/7/76 - 14/7/77	Contract No: 022-76 ESD Project No:

Contractor: Universität Würzburg

Address: Institut f. Biochemie der Universität, Röntgenring 11,
           D - 8700 Würzburg, Germany

Head of Project: Prof. Dr. D. Oesterhelt

Description of research work

I. Objectives (aims)

The retinal protein complex Rhodopsin serves a light sensoring function in the eye of higher animals. Its bacterial analogue bacteriorhodopsin mediates light energy conversion via its function as a light driven proton pump. In both cases retinal changes its spectral and chemical properties upon association with the protein. The chromophore formed plays the central role in the function of both chromoproteins. However, its chemical structure and structural changes following light absorption are unknown. The aim of the project is to elucidate the chromophore structure of the two retinal protein complexes in a comparative way.

2. Work Programme

The following basic experimental approaches to the elucidation of chromphore structure and function are feasible.

(i)   Equilibration of native chromophores with derivatives having different chemical reactivity. This can be achieved by use of organic solvent or photosteady state mixtures of chromophores produced in light;

(ii)   Removal of retinal from the binding site and reconstitution of the chromophore. The characterization of the intermediates of reconstitution gives insight into the retinal protein interaction;

                                       ../..

Total cost:
    DM.   104,000
    u.a.   28,415
	E. C. Contribution: 50%
	    DM. 52,000
	    u.a. 14,208

COMMISSION OF THE EUROPEAN COMMUNITIES ENERGY R & D PROGRAMME Objective: Solar Energy	Project or Sector: Photochemical, photoelectrochemical and photobiological processes

2. Work Programme cont/..

(iii) Reconstitution of the chromophore with retinal analogues;

(iv) Reconstitution of the chromophore with a chemically modified protein moiety;

(v) Correlation of chromophore formation and functional capacity of bacteriorhodopsin in intact Halobacterial cells or lipid vesicles.

The comparison of rhodopsin and bacteriorhodopsin will be possible with respect to characterization of the intermediates and reconstitution of the chromophores with retinal analogues.

3. Status

The results so far obtained concern the bacteriorhodopsin chromophore.

(a) In organic solvents like dimethylsufoxide the chromophore (purple complex λ_{max} 560 nm) which does not react with hydroxylamine equilibrates with a 500 nm chromophore. This species reacts stereospecifically with its retinal moiety under formation of syn-retinal oxime. The retinal moiety in this chromophore is bound to the $\mathcal{E}$-amino group of a lysine residue of the polypeptide chain as shown by mass-spectrometry of the isolated retinyl amino acid after reduction and hydrolysis of the protein. The 500 nm chromophore is photochemically active as is the native purple complex. However, preliminary results with bacteriorhodopsin containing lipidvesicles indicate that a proton pumping activity is not carried by this chromophore.

(b) Absorption of light causes a series of spectroscopic changes which compose the photochemical cycle of the purple complex. Under constant illumination the retinal moiety becomes reduced by borohydride. Since this reaction does not occur in the dark apparently one of the intermediate chromophores is a reducible species. The reduction product shows unusual spectroscopic properties which results from the specific interaction of the cyclohexene ring moiety of the retinal molecule with the protein. Chemical analysis excludes a retro-retinyl protein as the reduction product.

../..

COMMISSION OF THE EUROPEAN COMMUNITIES ENERGY R & D PROGRAMME Objective: Solar Energy	Project or Sector: Photochemical, photoelectrochemical and photobiological processes

3. Status cont/..

(c) From spectroscopic and chemical analysis of the intermediates of the chromophore reconstitution reaction we can conclude that the first step upon binding of retinal into the binding site is a distortion of the cylohexene ring by interaction with amino acid side chains of the protein.

Our further experiments deal with the synthesis of retinal analogues modified in the cyclohexene ring moiety in order to check the influence of the modification on specific binding, chromophore reconstitution and function of the reconstituted chromophores in photophosphorylation of the intact bacterial cell.

In the case of rhodopsin the intermediates of reconstitution will be compared with those of bacteriorhodopsin and the retinal analogues used for a comparison of the properties of the binding sites in the two retinal proteins.

COMMISSION OF THE EUROPEAN COMMUNITIES ENERGY R & D PROGRAMME Objective: Solar Energy	Project or Sector: Photochemical, photoelectrochemical and photobiological processes

Title:
Structural aspects of antennae function and of the primary charge separation in photosynthesis of green plants.

Duration: 12 months Period: 1/1/76 - 31/12/76	Contract No: 023-76 ESD Project No:

Contractor: Technische Universität Berlin

Address: Strasse des 17 Juni 135 (PC 14)
D-1000 Berlin 12, Germany

Head of Project: Prof. Dr. W. Junge

Description of research work

I. Objectives (aims)

There is only limited knowledge about the internal structure of the antennae and of the photochemical reaction centres of green plants. The principle aim of this project (four years) is to unravel the mutual orientation of antennae pigments and of the chemically active components within the reaction centres.

The aims in particular:

A. Antennae

(i) Relative orientation of the antennae pigments serving photosystems I and II, respectively;

(ii) Flexibility of this orientation (rotational diffusion of pigment-protein complexes);

(iii) Physiological relevance of this flexibility (correlation with the variability of spill over of energy between the photosystems);

B. Reaction Centres

(i) The internal structure of the photochemically active chlorophyll-a dimer in photosystem I (and later in photosystem II);

(ii) Optical identification of the chemical nature of the primary electron acceptor and its orientation with respect to the dimer.

../..

Total cost: DM. 101,500 u.a. 27,732	E.C. Contribution: 50% DM. 50,750 u.a. 13,866

COMMISSION OF THE EUROPEAN COMMUNITIES ENERGY R & D PROGRAMME Objective: Solar Energy	Project or Sector: Photochemical, photoelectrochemical and photobiological processes

2. Work Programme

Preparations of reaction centre particles from green plants are used at first.   Starting from particles with minimum chlorophyll contents per photochemical reaction centre the mutual orientation of chlorophylls and carotenoids which serve as antennae with respect to the special chlorophylls which undergo chemical reactions is analysed by a photoselection technique.

Immobilized particles are excited with linearly polarized light.   The resulting absorption changes are measured with light polarized perpendicular and parallel to the exciting one.   The linear dichroism of the absorption changes is evaluated to yield information on the mutual orientation of the excited and the observed transition moments.

After completion of the work with the less complicated reaction centres, these studies will be extended to the more complex intact chloroplast.

3. Status

Studies on six-polypeptide-reaction-centre-I-particles (Bengis & Nelson 1975) with 100 and 40 chlorophylls per centre revealed the following:

(a) The chlorophyll-a molecules within the centre, which absorb at 698 nm like the dimer, behave like a circularly degenerate system.   As the number of molecules is unknown (the dimer only - or the dimer plus very few chlorophylls with antennae function) this leaves room for two alternative interpretations: either the dimer is composed of two chlorophyll-a with their Q_y-axes almost perpendicular to each other - or if these are in parallel it is associated with another pair of chlorophyll-a with the Q_y-axes perpendicular to the ones of the dimer.   This is a first clue for a discrimination between proposed models for the dimer structure.

(b) The B-carotenes are arranged with their long axis in parallel to the dimer plane.   This is most favourable for their protective action against destructive photooxidation of antennae at the red end of the spectrum.

COMMISSION OF THE EUROPEAN COMMUNITIES ENERGY R & D PROGRAMME Objective: Solar Energy	Project or Sector: Photochemical, photoelectrochemical and photobiological processes

Title:
Refined analysis of the action of the electrical field in the membrane of photosynthesis.

Duration: 12 months Period: 13/9/76 12/9/77	Contract No: 024-76 ESD Project No:

Contractor: Technische Universität Berlin

Address: Strasse des 17. Juni, 135, 1000 Berlin 12, Germany

Head of Project: Prof. Dr. H.T. Witt

Description of research work

I. Objectives (aims)

The photosynthetic electron transport is energized by two light driven redox reactions which occur vectorial across the thylakoid membrane. Thereby, a transmembrane electric field is generated.

In this programme we investigate the structural arrangement of the pigments within the membrane and the influence of the transmembrane electric field on partial reactions in the chloroplast membrane.

2. Work Programme

(a) From the comparison of electrochromic absorption spectra in vitro and in vivo investigations on the structural orientation of the pigments within the chloroplast membrane are at work;

(b) It was clarified that from measurements of electrochromic absorption changes information can be obtained on changes of surface potentials;

(c) The action of external electric fields on partial reactions within the chloroplast membrane is under investigation.

3. Status

(i) New refined measurements of electrochromic absorption changes in vitro lead to an almost complete coincidence with those induced by light in vivo. It is concluded that in the average the long axis of the carotenoids forms an angle of 74° with the direction of the

../..

Total cost:
    DM. 104,000
    u.a.   28,415
	E. C. Contribution: 50%
	    DM. 52,000
	    u.a. 14,208

COMMISSION OF THE EUROPEAN COMMUNITIES ENERGY R & D PROGRAMME Objective:　　Solar Energy	Project or Sector: Photochemical, photoelectrochemical and photobiological processes

3. Status cont/..

(i) cont/..

electric field.　Conclusions are drawn concerning the preferential orientation of the dipole moment differences of the red and the blue absorption bands of the bulk chlorophylls in the membrane;

(ii) Experiments for detection of surface potentials as a function of pH have been started;

(iii) By application of an external field a transmembrane field is generated across the thylakoid membrane which is as high as that induced by light.　In this way in the dark ATP- synthesis can be observed with yields as high as light-induced ones.

COMMISSION OF THE EUROPEAN COMMUNITIES ENERGY R & D PROGRAMME Objective: Solar Energy	Project or Sector: Photochemical, photoelectrochemical and photobiological processes

Title:
   Fundamental Studies of Photochemical and Photobiological Systems for the use of solar energy.

Duration: 12 months Period: 1/11/76 - 31/10/77	Contract No: 025-76 ESD Project No:

Contractor: Fritz-Haber-Institut der Max-Planck-Gesellschaft

Address: 1 Berlin 33, Faradayweg 4-6, Germany

Head of Project: Prof. Dr. H. Gerischer

Description of research work

I. Objectives (aims)

The aim of the research project is to find stable photoactive materials for electrochemical solar cells. The obtainable efficiency for solar energy conversion shall be investigated. Since the composition of the electrolyte has great influence on the photodecomposition reactions which limit the stability, different solutions shall be studied. The search for means to improve the stability by a modification of the surface properties is another important field of investigation.

2. Work Programme

The photoelectrochemical response of semiconducting materials like CdS, CdSe, GaP and MoS_2 is studied. Materials are used in form of single crystals or as sputtered polycrystalline layers. Other ways of thin film production are also investigated. The stability of these materials against photodecomposition is checked. The effect of protective coatings is analysed. Different redox systems are studied with respect to the rate of electron transfer at the contact with such semiconductors.

3. Status

The stability problems of semiconductor electrodes have been analysed theoretically on the basis of thermodynamic and kinetic properties. Studies with CdS-electrodes in solutions of Fe^{3+}/Fe^{2+} - and S^{2-}/S_2^{2-} redox systems have demonstrated energy conversion efficiencies between 3 - 6% for solar radiation. Other systems are under preparation.

Total cost:
  DM. 118,950
  u.a. 32,500
	E.C. Contribution: 50%
	  DM. 59,475
	  u.a. 16,250

COMMISSION OF THE EUROPEAN COMMUNITIES ENERGY R & D PROGRAMME Objective: Solar Energy	Project or Sector: Photochemical, photoelectrochemical and photobiological processes

Title:

Construction of a biophotolytic reactor for the production of hydrogen from water, using solar energy.

Duration: 12 months Period: 3/12/76 - 2/12/77	Contract No: 026-76 ESN Project No:

Contractor: Universiteit van Amsterdam

Address: Laboratory of Biochemistry, B.C.P. Jansen Institute, Plantage Muidergracht 12, Amsterdam, The Netherlands

Head of Project: Professor Dr. E.C. Slater

Description of research work

I. Objectives (aims)

In principle it is possible to construct a biophotolytic reactor for the production of hydrogen from water, making use of solar energy.   The overall reaction sequence in such a reactor may be formulated by the following scheme

$$H_2O \xrightarrow{h\nu} \begin{array}{c} \frac{1}{2}O_2 \\ 2e \\ 2H^+ \end{array} \begin{array}{c} \text{hydrogenase} \\ \text{hydrogenase} \cdot H_2 \\ 2H^+ \end{array} \longrightarrow H_2$$

The photochemical splitting of water supplies electrons to the enzyme hydrogenase, which is able to catalyse the reduction of protons to hydrogen gas.

Indeed, certain micro-organisms produce hydrogen from water by this reaction sequence.   This is, however, essentially a side reaction, the physiological function of which is to dispose of excess reducing equivalents.   In order to be able to construct a hydrogen gas plant on a large scale it will be necessary either to isolate the enzymes required for the above reaction sequence on a large scale and stabilize them, or to synthesize stable catalysts that ape the natural ones.   A pre-requisite for a programme in this direction is to understand more about the mechanism of the action of the natural catalysts.   The research programme centres on hydrogenase, about which surprisingly little is known.

Total cost:
    Fl.   122,646
    u.a.   33,880
	E.C. Contribution: 50%
	    Fl. 61,323
	    u.a. 16,940

COMMISSION OF THE EUROPEAN COMMUNITIES ENERGY R & D PROGRAMME Objective: Solar Energy	Project or Sector: Photochemical, photoelectrochemical and photobiological processes

2. Work Programme

Extraction, purification and study of the properties and mechanism of action of hydrogenase from two types of phototrophic bacteria: the sulphur-purple bacterium Chromatium vinosum and the non-sulphur bacterium Rhodospirillum rubrum. Both bacteria can grow in the light under semi-anaerobic conditions and contain a hydrogenase that is reversibly deactivated by O_2, in contrast to hydrogenases from strict anaerobic organisms which are irreversibly inactivated by O_2. Since the final goal is a practical one, attention will be paid also to the isolation procedure, i.e., all steps must be suitable for large-scale preparations. The purification of hydrogenase from Chromatium vinosum has been described earlier (Gitlitz, P.H. and Krasna, A. I. (1975) Biochemistry 14, 2561-2567) but this procedure is not suited for large-scale isolations. There are no reports in literature on the purification of hydrogenase from Rhodospirillum rubrum. When sufficient amounts of pure hydrogenase have been accumulated, an intensive study by electron paramagnetic resonance spectrometry of the iron-sulphur cluster(s) in the protein and the effect of hydrogen will be undertaken.

3. Status

Batches (60 l) of cell cultures are now grown in a plexiglass vessel, with a high surface to volume ratio, illuminated with fluorescent lamps. Cells are ground with sand in mechanical mortars. After spinning down the cell walls, all the hydrogenase activity is solubilized with detergent (Triton X-114).

The R. rubrum enzyme is then collected by $\cdot(NH_4)_2SO_4$ fractionation and all the coloured material is removed by passing through a DEAE-cellulose column at rather high ionic strength. These steps give a colourless solution and a purification of 50-fold, based on H_2-uptake capacity, in 80% yield. The further purification of the enzyme by ion-exchange and adsorption chromatography and gel filtration is now in progress.

Since the hydrogenase of C. vinosum was found to be more heat stabile, a heating step (10 min, 65°C) was included before the $(NH_4)_2SO_4$ fractionation. After DEAE-cellulose chromatography, a purification of 100-200 times has been achieved. The further purification is in progress.

No special precautions, difficult to achieve on an industrial scale, are taken. All purification steps are carried out in air. The columns are at room temperature. At the present stage of purification cytochromes and other distinct chromophores are not detectable by optical spectroscopy. (Optical spectra of Fe-S clusters are rather weak and featureless). With EPR spectroscopy only a

../..

COMMISSION OF THE
EUROPEAN COMMUNITIES

ENERGY R & D PROGRAMME

Objective: Solar Energy

Project or Sector:

Photochemical, photoelectrochemical
and photobiological processes

3. Status cont/..

signal typical for a $(4Fe-4S)^{1-}$ cluster can be observed at
temperatures below 20 K in the oxidized enzymes.   The reduced
preparations show no signals indicating that there ferredoxin is
absent.

COMMISSION OF THE EUROPEAN COMMUNITIES ENERGY R & D PROGRAMME Objective: Solar Energy	Project or Sector: Photochemical, photoelectrochemical and photobiological processes

Title:

Study of mechanism and regulation of primary and associated reactions in photosynthesis.

Duration: 12 months Period: 13/9/76 - 12/9/77	Contract No: 027-76 ESN Project No:

Contractor: Rijksuniversiteit te Leiden

Address: Huygens Laboratory of the State University, Wassenaarseweg 78, Leiden, Netherlands

Head of Project: Dr. L.N.M. Duysens

Description of research work

I. Objectives (aims)

The ultimate objective of this research is to contribute to the elucidation of all possible pathways of the light-initiated electron transfer, the components and structures concerned, and the finding out of the reasons why these and not other pathways occur, and how the regulation of energy and electron transfer occurs. The acquired fundamental knowledge in this field is of great importance for applications of certain chemicals used for influencing the growth of plants. In the future this knowledge will be important in connection with selection procedures of plants (or artificial mutants therefrom), with a higher yield or with other useful properties, or for the engineering of artificial systems for conversion of light energy.

2. Work Programme

In purple bacteria the primary reaction after excitation of the photochemically active bacteriochlorophyll dimer P is the transfer within 10 ps of an electron from P to pheophytin, I. Reduced I, I^-, then transfers an electron in 0.2 ns to ubiquinone, X.

The following is our work programme. X is reduced by an external reductant, and by light absorption the reaction centre is brought in state $P^+ I^- X^-$. The various pathways of the back reaction $P^+ I^- \rightarrow P I$, which occurs in about 10 ns, will be studied by means of fluorescence, luminescence, and absorption spectroscopy and kinetics. One known pathway, efficient at low temperature, occurs via the triplet state $P^+ I^- \rightarrow P_T I \rightarrow P I$ + heat. Another possible pathway is $P^+ I^- \rightarrow P^* I \rightarrow$

../..

Total cost:
    Fl.   122,646
    u.a.   33,880
	E.C. Contribution: 50%
	    Fl. 61,323
	    u.a. 16,940

COMMISSION OF THE EUROPEAN COMMUNITIES ENERGY R & D PROGRAMME Objective: Solar Energy	Project or Sector: Photochemical, photoelectrochemical and photobiological processes

2. Work Programme cont/..

P I + luminescence.   In chromatophores and cells most of the luminescence (delayed*fluorescence) will be emitted by the excited antenna chlorophyll B*, generated by back transfer of energy: $P^+B \rightarrow P\ B^*$.

We are studying and are planning to study these reactions by measuring the kinetics of spectral fluorescence, luminescence and absorption changes down to the nanosecond time region, as a function of various parameters.   These experiments will be supplemented by experiments based on other techniques available in our laboratory, such as ESR measurements.

Similar experiments are planned on system 2 of algae.   Rapid luminescence experiments in our laboratory have indicated that an acceptor, W, different from the acceptor Q exists, possibly an intermediate between P680 and Q.   P680 or P_2 is a photoactive chlorophyll a dimer, Q a plastoquinone; P_2 and Q are analogous to P and X in purple bacteria.   Redox changes of P^+, I^- and possibly W can be studied by means of rapid absorption difference spectroscopy, as a function of parameters such as temperature, and certain compounds known to affect cell constituents involved in the reactions.

3. Status

One of our recent findings is that in certain purple bacteria "fluorescence" emission is inversely correlated with the yield of triplet formation as a function of temperature.   These experiments suggest that at room temperature the reaction $P^+\ I^-\ X \rightarrow P\ I\ X^-$ is an important pathway for the back reaction.   If that is correct, it should be possible to obtain from these experiments further important physico-chemical data about the primary reaction of purple bacteria.

COMMISSION OF THE EUROPEAN COMMUNITIES ENERGY R & D PROGRAMME Objective: Solar Energy	Project or Sector: Photochemical, photoelectrochemical and photobiological processes

Title:

Study of the photoelectrolysis of aqueous solutions on polymer membranes for hydrogen production.

Duration: 12 months Period: 17/6/76 - 16/6/77	Contract No: 028-76 ESB Project No:

Contractor: Université Libre de Bruxelles (ULB)

Address: Av. F.D. Roosevelt, 50
1050 Bruxelles, Belgium

Head of Project: E. Vander Donckt

Description of research work

I. Objectives (aims)

The ultimate aim of the project is to study the photoelectrolysis of aqueous solutions on polymer membranes with a view to produce hydrogen.

2. Work Programme

The experiments are started with films made of tetracyanoquinodimethane salts of N-alkylpolypyridinium. The conductivities are measured under dark and irradiated conditions as a function of temperature. The system showing a suitable variation of conductivity under irradiation will be selected for further studies.

It will then be attempted to induce an electric potential difference between the two sides of the film by creating an electric anisotropy. This will in the first approach be obtained by forming an electret.

3. Status

Electrets of the aforementioned polymers have been prepared and their electrical properties are under study.

Total cost:
  BF. 3,688,000
  u.a.   73,760
	E.C. Contribution: 25%
	  BF. 922,000
	  u.a. 18,440

COMMISSION OF THE EUROPEAN COMMUNITIES ENERGY R & D PROGRAMME Objective: Solar Energy	Project or Sector: Photochemical, photoelectrochemical and photobiological processes

Title:

    Genetic manipulation of photosystem I and II in chloroplast membranes.

Duration: 12 months Period: 21/6/76 - 20/6/77	Contract No: 029-76 ESDK Project No:

Contractor: United Breweries Limited

Address: Vesterfælledvej 100, DK-1799, Copenhagen V, Denmark

Head of Project: Diter von Wettstein, Dept. of Physiology, Carlsberg Laboratory

Description of research work

I.   Objectives (aims) and Work Programme

Genetic manipulation of a metabolic pathway and its product is used with a dual purpose.   Firstly, it achieves a detailed molecular character- ization of the pathway and its genetic as well as metabolic regulation. Secondly, it provides the genes and knowledge for a self-reproducing adaptation of a specific pathway to human requirements.   In the present case, the study is concerned with the conversion of solar energy into chemical energy by the photosynthetic membrane of a higher plant, barley. We have identified in barley over one hundred nuclear genes, which upon mutation give rise to defects in the organization or development of the photosynthetic membranes.   The following strategy is employed to find among these genes those which are specifically involved in the determin- ation of proteins in the thylakoid membrane.

Mutants that accumulate considerable amounts of chlorophyll and thylakoid membranes are tested for their ability to fix CO_2, and to evolve oxygen in vivo.   Mutants which are deficient in these capacities are tested for partial reactions of the photosynthetic electron transport chain.   The photosynthetic membranes of relevant mutants are isolated and the membrane polypeptides characterized by gelelectrophoresis and partial determination of their primary structure.   The organization of the mutant membranes is analyzed by freeze fracturing and freeze etching.   The possibility will be investigated of inserting certain wild type proteins into defective mutant chlrooplast membranes and thereby ascertain the function of the individual membrane polypeptides.

                                              ../..

Total cost:
    DK.   263,700
    u.a.   35,160
	E.C. Contribution: 50%
	    DK. 131,850
	    u.a. 17,580

COMMISSION OF THE EUROPEAN COMMUNITIES ENERGY R & D PROGRAMME Objective: Solar Energy	Project or Sector: Photochemical, photoelectrochemical and photobiological processes

2. Status

During the last year we have worked out a procedure to obtain pure internal membranes of young barley plastids.   Such young wild type plastids are more closely comparable to mutant plastids with respect to difficulty of isolation, purification and content of internal membranes.   A major obstacle in the analysis of plastid membranes of mutants has been the difficulty in obtaining high yields of uncontaminated membranes.   Results with mutant membranes of higher plants have so far been obtained by comparing contaminated mutant membranes with membranes from mature wild type chloroplasts, which are more easily obtained in pure form.

Prechilled leaves are homogenized in tricine buffer containing 0.6 M glycerol and calcium nitrate (using a Braun homogenizer fitted with replacable injector blades).   After filtration through two layers of nylon gauze the plastids are pelleted at 1400 x g.   The pellet is washed once and then shocked twice with 25 mM HEPES buffer containing 10 mM EDTA and the membranes are pelleted and resuspended in 1.90 M sucrose in 25 mM HEPES, 5 mM EDTA.   The membranes are purified by flotation through step-wise sucrose gradients of 1.30 M and 1.14 M sucrose using centrifugation for 1 hour at 140,000 x g.   The bank at the 1.30 M/1.14 M interface was shown by electron microscopy and gel electrophoresis to be free from nuclear and mitochondrial contamination. With the new isolation technique the polypeptide pattern of the non-photosynthetic barley etioplast membrane consists of 7 major bands and up to 10 minor bands.   Within three hours of greening, the membrane profiles contain 15 major and 35 minor bands and include all of the bands present in the etioplast pattern.   Apart from the disappearance of two of the major etioplast bands, only quantitative differences are observed during subsequent greening.

In co-operation with N.-H. Chua we have made an effort to correlate the barley thylakoid polypeptides with the 33 polypeptides found by him in Chlamydomonas.   Dr. Chua has raised monospecific antibodies against 9 of the Chlamydomonas polypeptides as well as antibodies against the total mixture of solubilized membrane polypeptides.   These were tested by two-dimensional immunoelectrophoresis against barley chloroplast and etioplast membrane polypeptides.   A considerable number (at least ten) of the antibodies in the serum against the Chlamydomonas polypeptide mixture reacted with wild type barley polypeptides.   Of the 9 monospecific antibodies, seven reacted with barley membrane proteins.   From the results it is apparent that seven major polypeptides in Chlamydomonas corresponded to seven major polypeptides in barley, although they differ in apparent molecular weight.

../..

COMMISSION OF THE EUROPEAN COMMUNITIES ENERGY R & D PROGRAMME Objective: Solar Energy	Project or Sector: Photochemical, photoelectrochemical and photobiological processes

Gel electrophoretic analysis showed that some major polypeptide bands of very young wild type plastids were due to contamination by nuclear proteins. These bands are also present in the polypeptide profiles of many of the mutant plastids and may now be considered as contaminants. This, however, does not invalidate the differences established between wild type and certain mutants.

Mutants viridis-c[12] lacks photosystem II activity and a membrane polypeptide of apparent molecular weight of 46,000 dalton. The latter protein of wild type barley membranes reacted with the antibody raised against the 47,000 dalton polypeptide of Chlamydomonas implicated by Chua and Bennoun to be the reaction centre protein of photosystem II_2. This antibody gave no reaction with thylakoid proteins of viridis-c[12]. This indicates that viridis-c[12] is comparable to some of the mutants in Chlamydomonas lacking photosystem II activity.

Interestingly, viridis-zd[69] - also deficient in photosystem II activity - contains the 46,000 dalton polypeptide, but has reduced amounts of another polypeptide with higher electrophoretic mobility, which also is affected in viridis-c[12]. The giant grana mutant xantha-b[12] did not show a qualitative difference in the band pattern after membrane solubilization in SDS which could be equated with the pronounced difference previously established by solubilization with phenol acetic acid urea. The grana deficient mutant viridis-k[23] lacks 3 thylakoid polypeptides, two of which are components of chlorophyll protein complex II.

The temperature sensitive mutant tigrina-o[34] produces at 20° collapsed grana in addition to lycopene crystalloids. These thylakoid abnormalities are paralleled by the absence of polypeptides belongine to chlorophyll protein complexes I and II and coupling factor. These mutant membranes thus have a polypeptide composition apparently different from both etioplast and mature chloroplast membranes. Growing the mutant at 30° results in a normal chloroplast ultrastructure and polypeptide composition.

We have started to investigate the question, whether the frequencies and size distributions of particles observed by freeze fracturing of plastid membranes vary with differences in membrane polypeptide patterns. Mutant membranes are to be compared with those of wild type etioplasts and chloroplasts. It has been found that prolamellar body membranes cross-fracture, primary thylakoids in etioplast have particle frequencies comparable to the corresponding fracture faces of chloroplast stroma lamellae. The perforated thylakoids originating by the dispersal of the prolamellar body fracture in the plane of the membrane and have a particle frequency which is about half that of primary thylakoids.

Mutant plastids are generally more fragile than wild type plastids and mutant plant material is available in smaller quantities. This necessitates working out procedures for making protoplasts from seedlings and isolating plastids from these. For wild type seedlings of barley intact plastids have been obtained in very high yield. With mutant leaves plastid yield has been high, but most of them have lost their outer envelope.

COMMISSION OF THE EUROPEAN COMMUNITIES ENERGY R & D PROGRAMME Objective: Solar Energy	Project or Sector: Photochemical, photoelectrochemical and photobiological processes

Title:

A study of the structure and function of the chloroplast membranes.

Duration: 12 months Period: 13/12/76 - 12/12/77	Contract No: 030-76 ESI Project No:

Contractor: Università di Milano, Istituto di Scienze Botaniche

Address: via Festa del Perdono, 7, Milano, Italy

Head of Project: Prof. G. Forti

Description of research work

I. Objectives (aims)

A study of the factors affecting the stability of isolated chloroplast membranes, and the type of damage suffered by membranes stored under different conditions. This will hopefully lead to devising methods to prevent the damage, and preserve the photochemical activity of the membranes.

2. Work Programme

Study of the time-course of inactivation of chloroplast membranes after various treatments intended to stabilise them. The loss of the following activities is being studied: (a) Photosystem II activity (reaction: H_2O to ferricyanide); (b) Photosystem I + II activity (reaction: H_2O to methylviologen); (c) Photosystem I activity (reaction: ascorbate+DCIP to methylviologen). Furthermore, structural alterations of the membranes will be studied by various methods (changes of absorption spectrum, of fluorescence, etc.).

The various treatments include: (i) glutaraldehyde treatment; (ii) removal of coupling factor (which is known to produce a large increase in the rate of electron transport); (iii) storage under nitrogen atmosphere; (iv) storage in the presence of serum albumin (BSA); (v) combination of these treatments. The causes of the inactivation of the above-mentioned reactions will be studied.

../..

Total cost: Lit. 36,875,000 u.a. 59,000	E.C. Contribution: 50% Lit. 18,437,500 u.a. 29,500

COMMISSION OF THE EUROPEAN COMMUNITIES ENERGY R & D PROGRAMME Objective: Solar Energy	Project or Sector: Photochemical, photoelectrochemical and photobiological processes

3. Status

The experiments so far performed have shown the following: (1) BSA by itslef has some protective effect; (2) glutaraldehyde ensures preservation of the activity for several days, though at low level, both in control and CF_1-deprived chloroplasts. CF_1-deprived chloroplasts, treated for 5 min. with glutaraldehyde then stored at room temperature in the presence of BSA (2 mg/ml) had no loss of activity for 88 hours. However, the treatment immediately lowered the activity to the basal level (120 electron equivalents hr^{-1}.mg chlorophyll), while the activity of the "uncoupled" control was 400 el. equiv. So, glutaraldehyde suppress the electron transport stimulation caused by removal of CF_1. In other words, this reagent acts as an "energy transfer inhibitor" even in membranes CF_1-less.

COMMISSION OF THE EUROPEAN COMMUNITIES ENERGY R & D PROGRAMME Objective: Solar Energy	Project or Sector: Photochemical, photoelectrochemical and photobiological processes

Title:

Photochemical conversion of solar energy by means of non-biological systems involving co-ordination compounds.

Duration: 12 months Period: 1/11/76 - 31/10/77	Contract No: 031-76 ESI Project No:

Contractor: Università di Bologna

Address: Via Selmi, 2 - 40126 Bologna
                        Italy

Head of Project: Prof. Luca Moggi

Description of research work

I. Objectives (aims)

In order to obtain an efficient conversion of the solar radiation, endothermic photoreactions must be included in appropriate, cyclic "catalysed" systems.   The purpose of this research is to study the properties of transition-metal coordination compounds for establishing whether they can be used as "catalysts" in cyclic photochemical systems.

2. Work Programme

Most of the endothermic photoreactions (and in particular, the photodissociation of water) being oxidation-reduction processes, the research will mainly be devoted to the evaluation of the redox properties of the excited states of coordination compounds: their oxidation and reduction potentials and the kinetic parameters of their redox reactions.   These data will be compared with the redox properties of ground-state complexes.   Attempts will also be made to rationalize them in terms of their dependence on the orbital nature of the excited state, the nature of the metal and/or ligands, the geometry of the complex, and other factors.

3. Status

Two lines of research are in progress:

(a)   study of bimolecular electron transfer processes of excited $Os(dipy)_3^{2+}$, $Ir(phen)_2Cl_2^+$, and $Ir(Me_2phen)_2Cl_2^+$ with homogeneous series of inorganic and organic compounds.

(b)   study of the photochemical and photophysical behaviour of $Cr(dipy)_3^{3+}$.

Total cost:
 	
    Lit.   36,875,000
    u.a.   59,000
	E.C. Contribution: 50%
	
	    Lit. 18,437,500
	    u.a. 29,500

Project E

Photosynthetic production of organic matter

COMMISSION OF THE EUROPEAN COMMUNITIES ENERGY R & D PROGRAMME Objective: Solar Energy	Project or Sector: Photosynthetic production of organic matter

Title: The utilization of excess straw for heating.	

Duration: 9 months Period: 1/8/76 - 30/4/77	Contract No: 051-76 ES DK Project No:

Contractor: Jordbrugsteknisk Institut Address: DK 2630 Tastrup, Denmark Head of Project: M. T. Tougaad Pedersen	

Description of research work

I. Objectives (aims)

Every year about 25% of the total straw production in Denmark is burned in the field.   The objective of the first phase of the project is to examine the possibilities for utilizing this excess of straw and give an economic evaluation of these possibilities.   It is also the aim to set up a research project concerning the use of excess straw for heating purposes.

2. Work Programme

The work programme for this study has been an investigation of the following points:

(i)   Straw production in Denmark;

(ii)   Mulching of straw;

(iii)   Improving the feeding value; (a) NaOH treatment; (b) NH_3 treatment;

(iv)   Production of paper and cardboard;

(v)   Production of straw particle board;

(vi)   Mixing of straw with beet silage;

(vii)   Other possibilities: (a) single cell production; (b) pyrolysis; (c) fermentation;

                                                        ../..

Total cost:
    DK.   75,000
    u.a. 10,000
	E. C. Contribution: 100%
	    DK. 75,000
	    u.a. 10,000

COMMISSION OF THE EUROPEAN COMMUNITIES ENERGY R & D PROGRAMME Objective: Solar Energy	Project or Sector: Photosynthetic production of organic matter

2. Work Programme cont/...

(viii)   Use of straw for heating purposes;

(ix)   Economic evaluation of the above-mentioned possibilities:

(x)   Proposal of a reaearch project concerning the use of excess for straw heating purposes.

3. Status

Work in progress.

COMMISSION OF THE EUROPEAN COMMUNITIES ENERGY R & D PROGRAMME Objective: Solar Energy	Project or Sector: Photosynthetic production of organic matter

Title:

Production of energy from straw (ascertainment of
the quantity of straw, determination of calorific value,
combustional systems, methods of hauling, transport and
storage of straw).

Duration: 12 months Period: 13/9/76 - 12/9/77	Contract No: 052-76 ESD Project No:

Contractor: Bayerische Landesanstalt für Landtechnik
Techn. Universität München

Address: D 8050 Freising
Federal Republic of Germany

Head of Project: Dr. Arno Strehler, Landtechnik Weihenstephan

Description of research work

I. Objectives (aims)

- Ascertainment of the actual amount of straw available in Germany
for energy consumption (at present and in the future);

- Assessment of the situation at present;

- Futuristic assessment;

- Further ascertainments of the calorific value of straw in consideration
of varieties and moisture content;

- Assessment of optimal systems of combustion in consideration of
capacity requirements;

- Assessment of optimal hauling, transport and storage methods for
straw in consideration of capacity and costs.

2. Work Programme

(a) Inquiries were placed at agricultural advisory boards and
ministeries and at national statistic boards.   From statistical
data on yields of grain, it is possible to establish data on
total yields of straw in consideration of the relationship between
grain and straw.   Special local situations, due to variations in
climate, and possibilities of using straw shall be determined.

Total cost:
    DM.   146,400
    u.a.   40,000
	E.C. Contribution: 50%
	    DM. 73,200
	    u.a. 20,000

COMMISSION OF THE EUROPEAN COMMUNITIES ENERGY R & D PROGRAMME Objective: Solar Energy	Project or Sector: Photosynthetic production of organic matter

2. <u>Work Programme</u> cont/...

(b) A series of examinations, refering to the calorific value with the method of isothermal water jacket, shall be continued. The influence of storage on growing conditions and on fertilizing shall also be investigated;

(c) Examination of straw furnaces according to the method of DIN 4702, further development of existing prototypes, measurement of labour consumption for handling straw furnaces;

(d) Measurement of capacity for various systems of hauling straw, determination of power requirements, labour consumption and costs. Development of "do-it-yourself methods" in order to erect inexpensive straw stores. Valuation of single systems in hauling, transporting and storing straw in consideration of economic aspects and of the necessity of power in relation to straw masses.

3. <u>Status</u>

(i) The quantities of straw produced annually were estimated for various regions in Bavaria. The present average consumption in Bavaria has already been estimated. These activities shall be expanded throughout the Federal Republic of Germany, final results are not yet at our disposal.

(ii) The calorific value of straw, in consideration of type, variety and moisture content, are at our disposal. The consideration of storage, fertilizing and climatic conditions during the growing period are being researched at present.

(iii) A testing place for the measurement of furnaces has already been erected, various plants have already been tested. Further tests are to be done with the aim of improving the furnaces. With the erection of furnaces for big bales on farms, together with the use of heatstorage, it will be possible to assess this system of straw combustion.

(iv) Assessment of methods for hauling, transporting and storing straw in consideration of costs and capacity has been done. Some results are already available and show that, in particular, the use of the big bale system has important labour saving advantages. Some new systems for erecting straw stores were developed and built up in practice. Some methods of covering straw with sheets were tested, new methods are at present being examined.

COMMISSION OF THE EUROPEAN COMMUNITIES ENERGY R & D PROGRAMME Objective: Solar Energy	Project or Sector: Photosynthetic production of organic matter

Title:

        Five studies on the production of straw and of maize stalks.

Duration: 11 months Period: 1/9/76 - 31/7/77	Contract No: 053-76 ESF Project No:

Contractor: Institut National de la Recherche Agronomique

Address: 149, rue de Grenelle, 75341 Paris, France

Head of Project: M.P. Chartier

Description of research work

I. Objectives (aims)

The biological conversion of solar energy is a process which has long been used in agriculture, forestry and in harvesting in a marine environment. Part of the biomass produced is at present not being used; a case in point being a considerable quantity of cereal straws and maize stalks which could be used to provide the energy equivalent of over 2,000,000 tonnes of petroleum in France today.

The aim of this study is to determine the quantities of material capable of recovery, taking into account the requirements for animal farming and for humus, together with the costs of recovery in terms of both energy and finance.

2. Work Programme

INRA will submit five reports on the follow subjects:

1. Characterization of the potentialities of the medium (soil, climate) with regard to production;

2. Examination of the efficiency of cereal and maize crops. Influence of cultivation techniques (nitrogenous fertilizer, date of sowing, seeding, crop health) and of genetic engineering on the straw yield;

3. Protection of soils being used for maximum yields of organic material. Assessment of the results of long-term experiments on the influence of straw, on the fertility of soils and on the structural
                                    ../..

Total cost:
   FF.   666,480
   u.a. 120,000
	E.C. Contribution: 42%
	   FF. 277,700
	   u.a. 50,000

COMMISSION OF THE EUROPEAN COMMUNITIES ENERGY R & D PROGRAMME Objective: Solar Energy	Project or Sector: Photosynthetic production of organic matter

2. Work Programme cont/..

qualities.

4. Energy assessment for the main species of cereals: input (fossil calories), output (food calories and straw calories). Variation in relation to cultivation systems and level of intensification.

5. Economic assessment of straw gathering.

3. Status

Survey and data collection being completed.

COMMISSION OF THE EUROPEAN COMMUNITIES ENERGY R & D PROGRAMME Objective: Solar Energy	Project or Sector: Photosynthetic production of organic matter

Title:

Theoretical and practical investigation of the possibility of producing energy at an economic cost from Terrestrial Biomass.

Duration: 12 months Period: 24/11/76 - 23/11/77	Contract No: 055-76 ES EIR Project No:

Contractor: An Foras Taluntais

Address: 19 Sandymount Ave., Ballsbridge, Dublin 4, Ireland

Head of Project:

Description of research work

I. Objectives (aims)

(a) Scan literature for yields of various species of herbaceous and woody plants appropriate to the climatic conditions of the EEC. Assess methods by which this biomass could be converted to energy. Ascertain in detail yields of short term forest species used in U.S. wood fibre studies. Ascertain the likelihood of commonly growing species such as willow, alder, poplars, etc. giving an economic yield.

(b) Ascertain to what extent the more promising U.S. species grow in Europe, and determine the area of soils suitable for this enterprise in EEC countries. Make broad projections of the areas likely to become available for this type of cropping in the nine countries.

(c) Outline a number of alternative methods of cultivating one or two of the species mentioned above; this to include coppicing versus other methods of propagation, systems of harvesting, transport and storage. Estimate the cost per unit of the heat from the different systems.

(d) Make projections as to the extent to which yields could be increased by breeding new varieties, introducing species from abroad, changing the configuration etc. in order to obtain maximum utilization of solar energy.

(e) On the basis of the foregoing calculations, determine the maximum output of utilisable energy per hectare, and the proportion of the national needs which could be supplied in this way, for the nine countries. Eight species will be grown on 10 acre plots at 4 centres.

Total cost:
  £   65,986
  u.a. 158,366
	E.C. Contribution: 50%
	  £ 32,993
	  u.a. 79,183

COMMISSION OF THE EUROPEAN COMMUNITIES ENERGY R & D PROGRAMME Objective: Solar Energy	Project or Sector: Photosynthetic production of organic matter

2. Work Programme

(i)   Make a literature survey of Biomass production in the U.S., U.K. and continental Europe. Consult with appropriate experts in this field;

(ii) Study the physical and economic constraints in regard to Biomass production in Europe;

(iii) Test by means of field experiments ways of overcoming some of the constraints found in (ii) above.

3. Status

A study has been made of the research literature on short rotations in the U.S. and in a number of European countries. Sites in the U.S., Holland, Northern Ireland (etc.) have been visited. Experience obtained has been utilised in the selection of species for evaluation under this contract. Samples of twelve species grown from plants and cuttings have been planted in five locations representative of soil types likely to be available in the medium term.

COMMISSION OF THE EUROPEAN COMMUNITIES ENERGY R & D PROGRAMME Objective: Solar Energy	Project or Sector: Photosynthetic production of organic matter

Title: Investigation of the potential use of photosynthetic production of organic matter as an energy feedstock.	

Duration: 12 months Period: 25/4/77 - 24/4/78	Contract No: 192-77 ESUK Project No:

Contractor: University of Reading Address: Early Gate RG6 2AT Reading, United Kingdom Head of Project:	

Description of research work

   I. Objectives (aims)

To arrive at the identification of the most promising biological sources of energy.

2. Work Programme

The following stages are being followed:

(a) a systematic assessment of the available biological materials and their potential for producing usable energy;

(b) an assessment of the technologies for the utilisation of crops as energy feedstocks;

(c) a first selection, based on (a) and (b) above, of promising energy crops;

(d) comprehensive survey of the land resources likely to be available in the UK for producing energy crops;

(e) economic assessment and detailed energy input/output analysis of the biological materials selected under (c) using the necessary characteristics (location, soil type, topography, current use, etc.) of the land resources from the survey

(f) identification of the most promising biological sources of energy in the UK.

3. Status   Work in progress.

Total cost:
    £   7,337
    u.a.   17,608
	E.C. Contribution: 100%
	    £ 7,337
	    u.a. 17,608

PROPOSALS

UNDER NEGOTIATION

PROPOSALS UNDER NEGOTIATION

Project C: Photovoltaic conversion

1. Fundamental study of recombination in silicon solar cells
   with a highly doped substrate.

   Katholieke Universiteit Leuven K.U.L., Belgium

2. To set up a data collation centre concerning the use of
   silicon solar cells for terrestrial applications.

   Katholieke Universiteit Leuven K.U.L., Belgium

3. Study and realization of photovoltaic cells made with
   heterostructures of thin film III - V compounds deposited
   on oxydized aluminium sheets, using organo-metallic compounds.

   Centre National d'Etudes Spatiales CNES, France

4. Study of a method allowing the synthesis of pure organo-
   metallic compounds in view of fabricating semi-conductors
   by chemical vapour deposition.

   Institut National de Recherche Chimique Appliquée IRCHA,
   France

Project E: Photosynthetic production of organic matter

1. Solar energy conversion by the photosynthetic production of
   organic matter: theoretical studies investigating the
   feasibility of the production of plant matter specifically for
   energy purposes in European conditions.

   Consiglio Nazionale della Ricerca - CNR, Italy

Geothermal energy

By virtue of its geographical distribution and the quantities of
energy which could be tapped, the possible overall contribution
of geothermal energy towards meeting Europe's future energy
requirements is much smaller than that of solar energy, but it
will not be negligible on a local scale.   As far as Europe is
concerned, geothermal energy is exploited only in Italy (electricity
generation from steam) and in France (space heating with hot water).

In order to assess the real potential of this form of energy, an
inventory of possible sources must be drawn up for all EC countries.
This requires not only compilation of existing information but also
actual field work in many areas.   Exploration of geothermal fields
requires improved geophysics and geochemical techniques and methods
of interpretation.   Also, certain developments related to the
exploitation of both low enthalpy and high enthalpy resources need
to be undertaken.   A host of problems connected with the not yet
proven hot dry rock concept must be studied, although a somewhat
cautious approach might be recommended in this field.   In order to
cope with the aforementioned problems, the EC programme undertakes
R & D work in the following areas or projects:

A. Acquisition and collation of existing and new geothermal data

A major task undertaken in the framework of the programme is the
accumulation and comparison of all information on geothermal data
available in the nine Community countries (temperature distribution
in the sub-surface, temperature gradient, heat flux, etc.).   This
should make it possible in a second step to identify those areas
where reservoirs could possibly exist and to estimate the energy
content and the possible yield of some of the reservoirs.

The accomplishment of the above mentioned task essentially requires
the acquisition of much additional data for each particular area
of potential geothermal interest.   Therefore, quite some field work
in those areas is to be undertaken.

B. Improvement of exploration methods

Exploration methods used extensively for oil or ore exploration
could be used for detection of temperature anomalies.   For this
purpose appropriate adaptations are undertaken both of the equipment
and the interpretation methods.   New methods and equipment are
developed, mainly in the field of geophysical methods like electric
sounding, electromagnetic sounding, magnetic sounding and seismic
methods.   Geochemical methods are improved to predict more
reliably the temperature of the hot spring waters in their original
reservoirs.

C. Sources of hot water (low enthalpy)

The work carried out in the field of low enthalpy geothermal water (temperature up to 100°C) covers the following topics:

- models to describe the geothermal reservoirs in appropriate areas;

- testing of actual wells and reservoirs to check the validity of the models;

- study of problems of exploitation and management including reinjection, rate of cooling of the reservoir etc.

Practical aspects of geothermal energy utilization are studied for urban heating, industrial and agricultural purposes.

D. Steam sources (high enthalpy)

The objective of this part of the programme is to improve the background knowledge on exploitation of high enthalpy fields (high temperature water and steam), reservoir management and associated technologies.

The adequate description (models) of the reservoir required for reservoir management in order to predict its behaviour during exploitation, and in particular, the effect of reinjection is investigated.   To obtain adequate measurements, appropriate equipment is developed which has to be more reliable for high temperatures in wells than the actually existing devices.   Drilling and more specifically well equipment for high temperatures is being improved.

E. Hot dry rocks

In this part of the programme emphasis is placed on feasibility studies for the possible utilization of thermal energy contained in hot dry rocks.   The heat of these rocks has to be extracted by circulation of water through appropriate cracks (to be created artificially).   It is estimated that the process could be attractive only for rocks at temperatures higher than 200°C.

The programme initially concentrates on the feasibility of creating adequate fracturation at great depth and increasing the permeability by chemical leaching, or other methods, to produce large enough areas for heat transfer.

SUMMARY AND BREAKDOWN

OF FUNDING

Objective: Geothermal Energy

Project	Number of contracts (*)	Total cost u.a.	E.C. contribution u.a.	Number of proposals under negotiation
A	16	1,626,700	863,945	1
B	19	1,228,376	585,205	1
C	6	445,369	232,336	1
D	8	716,495	362,028	0
Total	49	4,016,940	2,043,514	3

(*)Signed both by the Commission and the Contractor or sent for signature to the Contractor.

Project A

Acquisition and collation of existing
and new geothermal data

COMMISSION OF THE EUROPEAN COMMUNITIES **ENERGY R & D PROGRAMME** **Objective:** Geothermal Energy	**Project or Sector:** Acquisition and collation of existing and new geothermal data

Title:

Exploration of the temperature field to great depths in the area of Urach, as well as the testing of geophysical and geochemical methods.

Duration: 12 months **Period:** 1/1/77 - 30/12/77	**Contract No:** 071-76 EGD **Project No:** G/A23(D)

Contractor: Niedersächsisches Landesamt für Bodenforschung

Address: Alfred-Bentz-Haus, Postfach 51 01 53
D 3 Hannover 51

Head of Project: Prof.Dr. Behrens; Dr. Berkthold; Prof.Dr. Hahn; Prof.Dr. Schneider; Dr. Prodehl; Prof.Dr. Althaus; Prof.Dr. Sauer.

Description of research work

I. Objectives (aims)

This project is divided into seven sub-projects, the details of which appear in the following pages.

Total cost: DM 967,115 u.a. 264,239	**E.C. Contribution:** 58% DM 565,015 u.a. 154,376

Project No. G/A23(D)   Part 1

Sub-title: Measurements of thermal conductivity.

Institute: Institut für Angewandte Geophysik
             Technische Universität Berlin

Leader: Prof. Dr. J. H. Behrens

Objective:

The objective of the programme is to develop two needle probes for
in situ measurements of thermal conductivity. One of the needles
will serve to measure the thermal conductivity at the bottom of the
borehole and will be used in shallow holes to a depth of 30 m.
The second one shall be used for measurements along the wall of the
borehole.   In both cases, the probe will have to be pushed into the
rock being measured by an appropriate mechanism.

Work Programme

- survey of literature;

- theoretical optimisation and design;

- tests on a laboratory scale;

- if necessary, optimisation of handling mechanisms, signal
  transmission and treatment, accuracy;

- testing in a borehole.

Status

Just started

Project No. C/123(D)   Part 2

Sub-title: Magnetic and magnetotelluric soundings.

Institute: Institut für Allgemeine und Angewandte Geophysik
               Universität München

Leader: Dr. A Berkthold

Objective:

To study the temperature anomaly in the Urach area by means of
the magnetic and magnetotelluric sounding methods.   The anomalies
which may be detected will be interpreted in terms of electrical
conductivity.   On the basis of other methods used in the same
area, a correlation should be established between variations in
conductivity and temperature variations.

Work Programme

Detection of electrical conductivity variations to depths greater
than 5 km. along a profile across the Urach area and parallel to
the Schwäbische Jura.

A study of the possible correlation to other results.

Status

Just started

Project No. C/A23(D)   Part 3

Sub-title: Determination of the Curie Surface.

Institute: Niedersächsisches Landesamt für Bodenforschung, Hannover

Leader: Prof. Dr. A. Hahn

Objective:

Determination of the shape and depth of the Curie transition surface
in the Urach and Rhine Graben areas.

Work Programme

Specimens of rocks from the Urach and Rhine Graben areas, comparable
to those probably existing in the basement, will be tested in a
laboratory for the determination of the Curie temperature.

The depth of the Curie Surface will be computed on the basis of
existing magnetic field data and by the use of two methods of
calculation.

Status

Just started

Project No. G/A23(D)   Part 4

Sub-title: Seismological investigations in the Urach area.

Institute: Institut für Geophysik, Universität Stuttgart

Leader: Prof. Dr. G. Schneider

Objective:

To investigate the applicability of seismology for the identification of geothermal areas.

Work Programme

Three mobile stations will be constructed. They will be used in conjunction with fixed seismic stations in the area of Stuttgart.

Status

Just started.

Sub-title: Seismic prospections in the Urach area.

Institute: Geophysikalisches Institut, Universität Karlsruhe

Leader: Dr. C. Prodehl, Dr. D. Emter

Objective:

To use seismic prospection to determine, in particular, the configuration of the crystalline basement in the Urach area.

Work Programme

(a) interpretation of existing data;

(b) measurements based on the explosions produced in the normal operation of quarries in the area;

(c) execution of special bursts at carefully selected locations.

Status

Just started.

Project No. G/A23(D)   Part 6

Sub-title: Geochemical investigations.

Institute: Institut für Mineralogie, Universität Karlsruhe

Leader: Prof. Dr. A. Althaus

Objective:

To study the dissolution in water of some of the major constituants of minerals (Na, K, Ca, SiO_2) as a function of temperature.

Work Programme

Reaction in autoclaves up to 100 bars (rock samples from boreholes in the Urach area). Analysis of the solutions. Evaluation of kinetics and equilibrium conditions.

Status

Just started.

Project No. C/A23(D)   Part 7

Sub-title: Hydrogeology

Institute: Geologisches Landesamt Baden-Württemberg, Freiburg

Leader: Prof. Dr. K. Sauer

Objective:

To compile all data of importance relating to hydrogeology in
the Urach area.

Status

Just started.

COMMISSION OF THE EUROPEAN COMMUNITIES ENERGY R & D PROGRAMME Objective: Geothermal Energy	Project or Sector: Acquisition and collation of existing and new geothermal data

Title:

Geothermal investigation in shallow observation wells

Duration: 12 months Period: 4/11/76 - 3/11/77	Contract No: C73-76 ECN Project No: G/A9(N)

Contractor: Netherlands Organisation for Applied Scientific Research (T.N.O.)

Address: P.O. Box 285
NL Delft

Head of Project: Dr. F. Walter

Description of research work

I. Objectives (aims)

To measure temperatures in existing drillholes to a depth of 350 m, and to construct isothermal maps.

2. Work Programme

(a) Development and construction of the probes and data recording system;

(b) Measurement in about 100 wells located in selected areas of the central Graben and adjoining areas. These measurements will be made mainly in observation wells;

(c) Analysis and interpretation of data;

(d) Construction of maps.

3. Status

Measurements have already been carried out in about 40 wells.

Total cost: Fl. 140,600 u.a. 38,840	E.C. Contribution: 40% Fl. 56,240 u.a. 15,536

COMMISSION OF THE EUROPEAN COMMUNITIES ENERGY R & D PROGRAMME **Objective:** Geothermal Energy	**Project or Sector:** Acquisition and collation of existing and new geothermal data

Title:

    Research programme which is the collation of existing geothermal data for the land area of the U.K.

Duration: 12 months **Period:** 1/7/76 - 30/6/77	**Contract No:** 074-76 EGUK **Project No:** G/A25(UK)

Contractor: Natural Environment Research Council

Address: 5, Princes Gate, South Kensington,
London SW7 1QN

Head of Project: Dr. Burley (Institute of Geological Sciences)

Description of research work

I. Objectives (aims)

To compile all existing data on temperature distribution in the U.K. subsurface and on the chemistry of groundwater.

2. Work Programme

(a) Temperature in boreholes:- tabulation of temperature data available from measurements in boreholes. Construction of maps of isotherms at various depths below ground level;

(b) Heat flux:- this will be determined from equilibrium temperature measurements and data on thermal conductivity;

(c) Geochemical data:- they will be tabulated for waters from boreholes and springs. After estimating the effect of surface water mixing, the temperature of the original reservoir will be calculated from the silica content, Na/K and Na/K/Ca ratios.

3. Status

A rather large amount of data has already been compiled.

Total cost:
  £ 19,684
  u.a. 47,242
	E.C. Contribution: 50%
	  £ 9,842
	  u.a. 23,621

COMMISSION OF THE EUROPEAN COMMUNITIES **ENERGY R & D PROGRAMME** **Objective:** Geothermal Energy	**Project or Sector:** Acquisition and collation of existing and new geothermal data

Title:

Compilation of the available information on the Massif Central in order to evaluate the geothermal potential of the area.

Duration: 12 months **Period:** 11/11/76 - 10/11/77	**Contract No:** 078-76 EGF **Project No:** G/A30(F)

Contractor: Bureau de Recherches Géologiques Minières (B.R.G.M. Paris)

Address: B.P. 6009
F 45018 Orleans Cédex

Head of Project: Mr. Varet (Service Géologique National B.R.G.M. Orleans)

Description of research work

I. Objectives (aims)

To analyse the information available and to draw conclusions on the most promising areas and the work they would require.

2. Work Programme

The work will involve literature reviews and detailed discussions with specialists who have worked in the area. It will cover all major aspects of geology, volcanology, geophysics, geochemistry and hydrology.

3. Status

The literature has already been extensively reviewed.

Total cost: FF. 256,800 u.a. 46,237	**E.C. Contribution:** 50% FF. 128,400 u.a. 23,118

COMMISSION OF THE EUROPEAN COMMUNITIES ENERGY R & D PROGRAMME **Objective:** Geothermal Energy	**Project or Sector:** Acquisition and collation of existing and new geothermal data

Title:

Investigation of geothermal potential of the U.K. by hydrogeological, structural and geophysical methods.

Duration: 12 months **Period:** 1/1/77 – 31/12/77	**Contract No:** 084-76 EGUK **Project No:** G/A26(UK)

Contractor: Natural Environment Research Council

Address: 5, Princes Gate, South Kensington,
London SW7 1QN

Head of Project: Dr. Burley (Institute of Geological Sciences)

Description of research work

I. Objectives (aims)

To survey all the information available, and to interprete it, in order to evaluate the geothermal potential of the U.K., and to outline the work required to increase the information available.

2. Work Programme

(a) Geophysics:– compilation of aeromagnetic gravity and seismic data available, more particularly for the major sedimentary basins and the major intrusive bodies. In the case of the sedimentary basins, information will be used especially to estimate the configuration of the surface of the crystalline basement. This will be related to information gathered from deep boreholes;

(b) Geochemical and hydrological:– a survey of the Bath/Bristol district. This will be a detailed study based on the information available on the geology and hydrogeology of the area. In addition, detailed analysis of waters will be carried out to determine mixing of deep and shallow waters, flow velocity and age of water, origin of waters and the extent of their interaction with the reservoir rocks. Temperatures of these reservoirs will be calculated from the chemical composition of the water.

(c) Modelling:– mathematical models will be prepared for the study of water circulation and heat flow in a sedimentary basin.

3. Status   Just started.

Total cost: £ 60,465 u.a. 145,116	**E.C. Contribution:** 54% £ 32,508 u.a. 78,019

COMMISSION OF THE EUROPEAN COMMUNITIES **ENERGY R & D PROGRAMME** **Objective:** Geothermal Energy	**Project or Sector:** Acquisition and collation of existing and new geothermal data

Title:

Survey of the geochemical data on Italian hot springs, evaluation of the geothermal reservoirs and reconstruction of the hydrogeology of some selected areas.

Duration: 9 months **Period:** 31.12.76 - 30.9.77	**Contract No:** 087-76 EGI **Project No:** G/A1(I)

Contractor: Istituto Internazionale per le Ricerche Geotermiche (C.N.R.)

Address: Lungarno Pacinotti, 55
I 56100 Pisa

Head of Project: Dr. Panichi

Description of research work

I. Objectives (aims)

The objective of this project is to collect all the data available on hot springs, and when necessary, to produce new analyses. On the basis of these analyses, the temperature of the water in the original reservoirs in the subsurface will be estimated and maps will be constructed.

2. Work Programme

(a) Investigation of the whole national territory

The data will be collected from literature and from public or private institutions. For the most important springs additional chemical analysis will be carried out. The temperatures will be calculated by the usual methods, based on silica content, or on the Na, K, Ca ratio. Calculations will be carried out by means of computer programmes to check the equilibrium of water with the most common minerals found in the possible reservoirs. Account will be taken of the fact that the ions are not necessarily free, but may be present as complexes. The temperatures obtained will be shown on maps at the scale of 1/1,000,000.

../..

Total cost: Lit. 73,345,000 u.a. 117,352	**E.C. Contribution:** 50% Lit. 36,672,000 u.a. 58,676

COMMISSION OF THE EUROPEAN COMMUNITIES	Project or Sector:
ENERGY R & D PROGRAMME	Acquisition and collation of existing and new geothermal data
Objective: Geothermal Energy	

2. Work Programme   cont....

(b)   Detailed study of the geothermal area of Siena

In addition to the collection of existing chemical analyses, samples will be taken of the most important hot springs of the area and will be analysed in order to:

- estimate the amount of mixing with water from the surface;

- evaluate the equilibria with minerals;

- outline the possible hydro-geological system.

This work will require also isotopic analysis, mainly of ^{18}O, deuterium, tritium and ^{14}C.

3. Status

Just started.

COMMISSION OF THE EUROPEAN COMMUNITIES ENERGY R & D PROGRAMME Objective: Geothermal Energy	Project or Sector: Acquisition and collation of existing and new geothermal data

Title:

Geothermal gradient and heat flux in Italy

Duration: 24 months Period: 1.4.77 - 31.3.79	Contract No: 088-76 EGI Project No: G/A27(I)

Contractor: Consiglio Nazionale delle Ricerche (C.N.R. Roma)

Address: Lungarno Pacinotti, 55
I 56100 Pisa

Head of Project: Dr. Fanelli (Istituto Internazionale per le Ricerche Geotermiche, Pisa)

Description of research work

I. Objectives (aims)

Compilation of temperature and heat flow data available.

2. Work Programme

(a) Temperatures:- for each location (boreholes, mines) the temperatures will be tabulated. Maps of isotherms will be constructed for various depths below ground level.

(b) Heat flux:- it will be calculated from geothermal gradient, thermal conductivity and the appropriate directions.

3. Status

Just started.

Total cost: Lit. 48,833,000 u.a. 78,133	E. C. Contribution: 50% Lit. 24,416,000 u.a. 39,066

COMMISSION OF THE EUROPEAN COMMUNITIES ENERGY R & D PROGRAMME Objective: Geothermal Energy	Project or Sector: Acquisition and collation of existing and new geothermal data

Title:

Study of the geothermal potential in the southern Appennin area.

Duration: 12 months Period:	Contract No: 090-76 EGI Project No: G/A2(I)

Contractor: Università degli Studi di Napoli
Istituto di Geologia e Geofisica

Address: Largo S. Marcellino, 10
I 80138 Napoli

Head of Project: Prof. B. d'Argenio

Description of research work

I. Objectives (aims)

This project will serve to collect all information connected with this large area and of possible relevance to geothermal energy. The data will be interpreted and an attempt will be made to indicate what possible areas would be of geothermal interest and what investigations could usefully be undertaken.

2. Work Programme

The following data will be collected:

(a) Geology:- all the necessary information will be put together in order to construct detailed maps of the geological structures. Particular importance will be given to the subsurface structures and the circulation of underground water;

(b) Volcanology;

(c) Thermal data:- temperatures and chemical analysis of hot springs;

(d) Hydrogeology;

(e) Geophysical data:- in particular, gravimetric and magnetic data as well as electrical sounding;

(f) Evaluation and interpretation.

3. Status - Being signed

Total cost: Lit. 44,100,000 u.a. 70,560	E.C. Contribution: 50% Lit. 22,050,000 u.a. 35,280

COMMISSION OF THE EUROPEAN COMMUNITIES ENERGY R & D PROGRAMME **Objective:** Geothermal Energy	**Project or Sector:** Acquisition and collation of existing and new geothermal data

Title:

Temperature distribution in the upper earth mantle under the Massif Central for the past five million years.

Duration: 12 months **Period:** 1/1/77 - 30/12/77	**Contract No:** 095-76 EGF **Project No:** G/A15(F)

Contractor: Institut National d'Astronomie et de Géophysique (CNRS)

Address: Dd Michelet 38
BP 1044, F 44037 Nantes Cédex

Head of Project: M. Nicolas (University of Nantes)

Description of research work

I. Objectives (aims)

To determine the origin of basalts and the depth of the earth mantle at the time of the eruptions.

2. Work Programme

(a) Collection of samples of spinel lherzolites nodules;

(b) Geochemical study of the rocks and in particular of the pyroxene minerals to obtain information on the temperature of the mantle at the time of the ejection of basalt.

(c) Study of the substructures due to deformation in order to obtain indications of the ambient pressures.

(d) Estimation of the age of the nodules.

3. Status

Just started.

Total cost: FF. 169,610 u.a. 30,538	**E.C. Contribution:** 47% FF. 80,000 u.a. 14,404

COMMISSION OF THE EUROPEAN COMMUNITIES ENERGY R & D PROGRAMME Objective: Geothermal Energy	Project or Sector: Acquisition and collation of existing and new geothermal data

Title:

U.K. geothermal exploration - heat flow studies.

Duration: 12 months Period: 1/1/77 - 31/12/77	Contract No: 096-76 EGUK Project No: G/A21(UK)

Contractor: University of Oxford

Address: Department of Geology and Mineralogy, Parks Road,
GB Oxford OX1 3PR

Head of Project: M. E. R. Oxburgh

Description of research work

I. Objectives (aims)

To produce new heat flow measurements for the territory of Great Britain.

2. Work Programme

Boreholes which have been drilled and abandoned by private companies and other organisations will be used by the contractor for these measurements. Only those among them with appropriate locations to provide really new information will be selected. The contractor will provide each borehole with a lining and a central tubing, which will then be sealed at the top. Temperature measurements will be carried out through the central tube until such time as thermal equilibrium is reached.

Thermal conductivity and radio-active element content will be measured on rock samples from the bottom of the borehole in order to determine the heat flux from the mantle.

3. Status

Just started.

Total cost: £ 40,145 u.a. 96,348	E.C. Contribution: 64% £ 25,840 u.a. 62,016

COMMISSION OF THE EUROPEAN COMMUNITIES ENERGY R & D PROGRAMME Objective: Geothermal Energy	Project or Sector: Acquisition and collation of existing and new geothermal data

Title:

The investigation of the S.W. England thermal anomaly zone.

Duration: 12 months Period: 1/9/76 - 31/8/77	Contract No: 097-76 EGUK Project No: G/A20(UK)

Contractor: Imperial College of Science and Technology

Address: Geophysics Dept., Prince Consort Road, GB London SW 7 2 BP

Head of Project: Dr. Wheildon

Description of research work

I. Objectives (aims)

To determine the heat flux from the earth mantle under the granite batholith of south-west England.

2. Work Programme

Boreholes will be drilled, initially in the Carnmenellis granite body, to a depth of 100 m. In each borehole the following measurements will be made:

- thermal conductivity measurements of rock samples and 'chips', and analyses to determine the heat production from radio-active elements;

- temperature measurements taken until thermal equilibrium is reached;

- determination of the heat flux coming from the mantle.

In the case of a positive thermal anomaly being confirmed, boreholes would also be drilled and investigated in neighbouring granite bodies in that area.

3. Status

Just started.

Total cost:
 	
  £ 21,351
  u.a. 51,242
	E.C. Contribution: 64%
	
	  £ 13,637
	  u.a. 32,720

COMMISSION OF THE EUROPEAN COMMUNITIES ENERGY R & D PROGRAMME **Objective:** Geothermal Energy	**Project or Sector:** Acquisition and collation of existing and new geothermal data

Title:

Reconstruction of the magma chamber under the Laacher volcano (East Eifel).

Duration: 12 months **Period:**	**Contract No:** 098-76 EGD **Project No:** G/A24(D)

Contractor: Ruhr Universität Bochum

Address: Universitätsstrasse 150, Postfach 2148
D 463 Bochum-Querenburg

Head of Project: Prof. Dr. H. U. Schmincke

Description of research work

I. Objectives (aims)

To estimate the shape and the volume of the magma chamber which existed under the volcano before its last eruption, 10,000 years ago. From a study of the erupted material it is expected to determine the amount of magma still left in the chamber, and to deduce whether there is still heat available at useful depths.

2. Work Programme

(a) Stratigraphic analysis of the pyroplastic deposits and determination of their volume;

(b) Chemical and mineralogical study of the pyroplastic deposits;

(c) Investigation of the nature and amount of xenoliths;

(d) Detailed investigation of the possible process and extent of magma differentiation.

3. Status

Being signed.

Total cost: DM. 165.720 u.a. 45.279	**E.C. Contribution:** 64,75 % DM. 107.300 u.a. 29.317

COMMISSION OF THE EUROPEAN COMMUNITIES ENERGY R & D PROGRAMME Objective: Geothermal Energy	Project or Sector: Acquisition and collation of existing and new geothermal data

Title:

Collection of geothermal data for Denmark

Duration: 12 months Period: 1/11/76 - 30/10/77	Contract No: 102-76 EGDK Project No: G/A4(DK)

Contractor: Aarhus Universitet

Address: Finlandsgade, 6
              D 8200   Aarhus N

Head of Project: Prof. Strend Saxov

Description of research work

I. Objectives (aims)

First stage survey of the geothermal potential of Denmark.

2. Work Programme

(a) Compilation of temperature data from fifty deep boreholes and the construction of maps of isotherms;

(b) Tabulation of all chemical analysis of water from boreholes and springs;

(c) Compilation of information relevant to hydrogeology (permeability of aquifers, pressure etc.);

(d) Interpretation of seismic and gravimetric data.   Map of the top surface of the Precambrian basement;

(e) Equipment for temperature measurements will be installed on the bottom of four lakes to record the temperature over a period of 1 year.

3. Status

    Just started

Total cost:
   DK. 132,880
   u.a. 17,718
	E. C. Contribution: 50%
	   DK. 66,440
	   u.a. 8,859

COMMISSION OF THE EUROPEAN COMMUNITIES ENERGY R & D PROGRAMME **Objective:** Geothermal Energy	**Project or Sector:** Acquisition and collation of existing and new geothermal data

Title:

        Compilation and analysis of the information on the Naples area and its geothermal potential. Brief survey of the Etna area and the Eolian Islands.

Duration: 12 months **Period:** 1/1/77 - 31/12/77	**Contract No:** 165-76 EGI **Project No:** G/A31(I)

Contractor: Osservatorio Vesuviano

Address: I 80056 Ercolano (Napoli)

Head of Project: Prof. Rapolla

Description of research work

I. Objectives (aims)

To compile all the information available on these areas and to interprete them from the point of view of geothermal energy.

2. Work Programme

- Temperatures and heat flux;

- Geological structures and volcanos;

- Geophysical data;

- Geochemistry and detailed petrographic data (in particular those resulting for example from fluid inclusions in minerals);

On the basis of this information the geothermal potential of each area will be discussed and the work necessary for a more detailed exploration of the most promising zones will be outlined.

3. Status

Just started.

Total cost:
    Lit. 34,000,000
    u.a.   54,400
	E.C. Contribution: 50%
	    Lit. 17,000,000
	    u.a. 27,200

COMMISSION OF THE EUROPEAN COMMUNITIES ENERGY R & D PROGRAMME Objective: Geothermal Energy	Project or Sector: Acquisition and collation of existing and new geothermal data

Title:

Compilation of geothermal data in France

Duration: 9 months Period: 1/1/77 - 30/9/77	Contract No: 170-76 EGF Project No: G/A5(F)

Contractor: Bureau de Recherches Géologiques et Minières (B.R.G.M. Paris)

Address: B.P. 6009
F 45018 Orleans Cédex

Head of Project: Mr. Lavigne / Mr. Mayer (B.R.G.M. Orleans)

Description of research work

I. Objectives (aims)

To compile all available geothermal data: temperature, heat flux and geochemical data.

2. Work Programme

(a) Temperatures:- the temperatures will come from boreholes, mines or hot springs. They will be tabulated for each location. They will be used to construct maps of isotherms at various levels to 3,000 m below ground level;

(b) Heat flux:- the heat flux data will be accompanied by the values of geothermal gradient and thermal conductivity;

(c) Geochemistry:- geochemical data will be collected for springs and boreholes. An attempt will be made to interpret them in order to evaluate the maximum temperature in the respective reservoirs.

3. Status

Just started.

Total cost: FF 288,720 u.a. 51,984	E.C. Contribution: 50% FF. 144,360 u.a. 25,992

COMMISSION OF THE EUROPEAN COMMUNITIES ENERGY R & D PROGRAMME Objective: Geothermal Energy	Project or Sector: Acquisition and collation of existing and new geothermal data

Title: Investigation of the geothermal anomaly of Urach with possible commerical explcitation in view.

Duration: 12 months Period:	Contract No: 176-77 EGD Project No: G/A22(D)

Contractor: Stadt Urach

Address: Freiburg I. BR.

Head of Project: Prof. Dr. Sauer (Geologisches Landesamt Baden-Württemberg)

Description of research work

I. Objectives (aims)

To drill a borehole in the area of Urach which should ultimately reach the crystalline basement at a depth which is established to be below 2,000 metres. For this contract a depth of only 1,600 m. will be reached.

2. Work Programme

(a) Drilling:- the initial diameter of the borehole will be 445 mm., to a depth of 470 m. It will be decreased by stages to 215 mm. at a depth of 1,600 m.;

(b) Geological observations:- when particular features are expected, cores will be drilled for inspection;

(c) Aquifers:- after drilling through aquifers, tests will be done to characterise them, after which appropriate additional hydraulic fracturation tests or chemical leaching tests will be made to stimulate permeability. Subsequently, they will be sealed by the injection of cement;

(d) Measurements:- measurements along the borehole will be carried out by a firm of specialists (induction electrical survey, density, sonic velocities etc.);

(e) Laboratory studies:- the cores will be studied in the laboratory, and in particular the following measurements will be made: permeability, thermoconductivity, petrographic examination.

../..

Total cost: 　DM.　1,725,591 　u.a.　471,472	E.C. Contribution:　50% 　DM.　862,796 　u.a.　235,736

COMMISSION OF THE EUROPEAN COMMUNITIES ENERGY R & D PROGRAMME **Objective:** Geothermal Energy	**Project or Sector:** Acquisition and collation of existing and new geothermal data

2. Status

Should be signed shortly.

Project B

Improvement of methods of exploration

COMMISSION OF THE EUROPEAN COMMUNITIES ENERGY R & D PROGRAMME **Objective:** Geothermal Energy	**Project or Sector:** Improvement of methods of exploration

Title:

    Exploration and interpretation of the high temperature gradient in the Rhine Graben.

Duration: 12 months **Period:** 4/11/76 - 3/11/77	**Contract No:** 075-76 EGD **Project No:** G/F14(D)

Contractor: Universität Karlsruhe

Address: Hertzstr. 16 - Bau 42
         D 75 Karlsruhe-West (21)

Head of Project: Prof. K. Fuchs (Geophysikalisches Institut, Universität Fridericiana Karlsruhe)

Description of research work

I. Objectives (aims)

To contribute to information on the possible causes of the thermal anomaly in the Rhine Graben.

2. Work Programme

(a) accurate temperature measurements as a function of depth, in boreholes which have reached equilibrium;

(b) interpretation of this new temperature data, together with all existing data; particular attention will be given to local irregularities of the temperature curves and the circulation of waters which could cause them;

(c) on the basis of this and other hydrogeological information, models will be constructed to interprete the observations and eventually conclusions reached as to whether the thermal anomaly of the Graben is of pure hydrological origin or has some other cause.

3. Status

Temperature measurements have been initiated.

Total cost:
    DM. 178,500
    u.a. 48,770
	E.C. Contribution: 56.86%
	    DM. 101,500
	    u.a. 27,732

COMMISSION OF THE EUROPEAN COMMUNITIES ENERGY R & D PROGRAMME Objective: Geothermal Energy	Project or Sector: Improvement of methods of exploration

Title:

Geochemical methods applied to the improvement of knowledge concerning geothermal reservoirs.

Duration: 12 months Period: 4/11/76 - 3/11/77	Contract No: 076-76 EGF Project No: G/B8(F)

Contractor: Université Louis Pasteur

Address: rue René Descartes, 5
F 67084 Strasbourg Cédex

Head of Project: Mr. Tardy

Description of research work

I. Objectives (aims)

Geochemical study of waters from boreholes in the Rhine Graben and from hot springs in the Vosges area.

2. Work Programme

The locations from which the samples will be taken will be selected on the basis of the geology being well-known and, therefore, where precise information can be obtained on the rocks with which underground waters are in contact. These rocks will be studied in particular on drilled cores available.

In addition to chemical analysis, the water samples will be submitted to isotope analysis both on dissolved substances and on gases.

Thermodynamic equilibria between waters and mineral species will be calculated as a function of temperature and pressure.

Temperatures will be calculated from SiO_2, Na-K and Na-K-Ca content. The possibility of using isotope ratios as geothermometers will be investigated.

3. Status

Just started.

Total cost: FF. 440,000 u.a. 79,222	E.C. Contribution: 50% FF. 220,000 u.a. 39,611

COMMISSION OF THE EUROPEAN COMMUNITIES ENERGY R & D PROGRAMME Objective: Geothermal Energy	Project or Sector: Improvement of methods of exploration

Title:

Study of the applicability of the d.c. electric sounding method for the geothermal exploration of the Rhine Graben.

Duration: 12 months Period: 11/11/76 - 10/11/77	Contract No: 079-76 EGF Project No: G/B7(F)

Contractor: Bureau de Recherches Géologiques et Minières (B.R.C.M. Paris)

Address: B.P. 6009
F 45018 Orleans Cédex

Head of Project: R. Horn / J. M. Georgel (Service Géologique National)

Description of research work

I. Objectives (aims)

To improve the electric sounding technique for geothermal exploration and to test it in the Rhine Graben.

2. Work Programme

The contractor will first adapt existing equipment for the use of large spacing Schlumberger type sounding and dipole-dipole type sounding. These methods will be used on a site in the Graben which is still to be selected. The experiments will include some variations of the main parameters in order to evaluate their influence on the quality of the measurements.

3. Status

Part of the work to modify the equipment is complete.

Total cost: FF. 287,243 u.a. 51,718	E.C. Contribution: 50% FF. 143,622 u.a. 25,859

COMMISSION OF THE EUROPEAN COMMUNITIES ENERGY R & D PROGRAMME **Objective:** Geothermal Energy	**Project or Sector:** Improvement of methods of exploration

Title:
Study of the conductivity anomalies in the Rhine Graben.

Duration: 12 months **Period:** 21/12/76 - 20/12/77	**Contract No:** 080-76 EGF **Project No:** G/B4(F)

Contractor: Institut National d'Astronomie et de Géophysique (C.N.R.S.)

Address: rue Lhomond, 24,
F 75231 Paris Cédex 05

Head of Project: Mr. J. Mosnier (Centre de Recherches Géophysique)

Description of research work

I. Objectives (aims)

To construct a map of the distribution of the value of conductivity in the French side of the Rhine Graben.

2. Work Programme

The measurements will be carried out by the differential magnetic sounding method. The contractor will initially use the existing equipment which covers a frequency range from 0.001 Hz to 0.1 Hz. In addition, some new equipment will be built to cover the range 0.01 Hz to 30 Hz and which should be able to sound structures to depths of 5 kilometres. The conclusions of these measurements will be compared with results of other methods to be used by teams in other areas.

3. Status

Just started.

Total cost: FF. 328,400 u.a. 59,129	**E.C. Contribution:** 50% FF. 164,200 u.a. 29,565

COMMISSION OF THE EUROPEAN COMMUNITIES ENERGY R & D PROGRAMME **Objective:** Geothermal Energy	**Project or Sector:** Improvement of methods of exploration

Title:

Development of a magnetic prospection method for the study of warm subsurface areas in the Rhine Graben. Determination of the depth of the Curie Surface.

Duration: 12 months **Period:** 1/8/76 - 30/7/77	**Contract No:** 082-76 EGF **Project No:** G/B3(F)

Contractor: Institut National d'Astronomie et de Géophysique (Paris)

Address: rue René Descartes, 5,
F 67084 Strasbourg Cédex

Head of Project: Prof. A. Roche (Institut de Physique du Globe, Strasbourg)

Description of research work

I. Objectives (aims)

To determine the depth and shape of the Curie Surface in the Rhine Graben area from Strasbourg to Karlsruhe and also in some neighbouring zones of the Vosges and Schwarzwald.

2. Work Programme

The investigation will attempt to take into account the real values of the magnetic properties of the rocks of the basement. Samples of representative rocks of the crystalline basement (granites, microgranites, diorites, etc.) will be taken from both sides of the Rhine Graben and studied in a laboratory. The following investigations will be made:

- measurement of magnetic susceptibility;

- measurement of magnetic viscosity;

- determination of the Curie point;

- Microscopic examination before and after heating.

Temperature measurements will be taken in existing boreholes, and the thermal conductivity of the rocks will be determined from these

../..

Total cost: FF. 561,141 u.a. 101,034	**E.C. Contribution:** 45.8% FF. 257.003 u.a. 46,303

COMMISSION OF THE EUROPEAN COMMUNITIES ENERGY R & D PROGRAMME Objective: Geothermal Energy	Project or Sector:

2. Work Programme cont...

boreholes on those samples which are available.

Magnetic field measurements will be made to add **weight** to those already available.

An attempt will be made during the first year of work to outline the isotherm in the area.

3. Status

Just started.

COMMISSION OF THE EUROPEAN COMMUNITIES ENERGY R & D PROGRAMME **Objective:** Geothermal Energy	**Project or Sector:** Improvement of methods of exploration

Title:

        Execution of vertical electric sounding of the dipole type in areas of geothermal interest.

Duration: 12 months **Period:** 1/1/77 - 30/12/77	**Contract No:** 021-76 EGI **Project No:** G/B34(I)

Contractor: Università degli Studi di Bari, Instituto di Geodesia e Geofisica

Address: I 70122 Bari.

Head of Project: Dr. Patella

Description of research work

I. Objectives (aims)

To investigate the applicability of this method for the determination of the vertical and horizontal extension of geothermal reservoirs.

2. Work Programme

The contractor will first determine the best technique of the dipole-dipole type to obtain the highest sensitivity and improve the method of interpretation of the signals. Mathematical models will be established in particular to simulate the heterogeneities which could correspond to the conductivity anomalies detected. The first measurements will be made in areas where the geothermal reservoir has already been well defined.

3. Status

Just started.

Total cost:
    Lit. 12,060,000
    u.a. 19,296
	E.C. Contribution: 60%
	    Lit. 7,236,000
	    u.a. 11,578

COMMISSION OF THE EUROPEAN COMMUNITIES ENERGY R & D PROGRAMME **Objective:** Geothermal Energy	**Project or Sector:** Improvement of methods of exploration

Title:

    Models for the interpretation of the results of electric soundings in non-tabular structures.

Duration: 12 months **Period:** 1/1/76 - 30/6/72	**Contract No:** 101-76 EGF **Project No:** G/B15(F)

Contractor: Institut National d'Astronomie et de Géophysique (C.N.R.S.)

Address: Place Eugène Bataillon,
F 34060 Montpellier Cédex

Head of Project: Mr. Vasseur (Université des Sciences et Techniques du Languedoc (CNRS/INAG)

Description of research work

I. Objectives (aims)

To develop a mathematical model for the interpretation of electric soundings in non-tabular structures, with particular emphasis on geothermal reservoirs.

2. Work Programme

A mathematical model will have to be developed for this kind of interpretation. It will then be used to calculate the effect of a simple heterogeneity of conductivity in a stratified structure. Several shapes will be considered (sphere, cylinder, disc, parallelepiped). The contractor will also consider several sizes and several depths at which the heterogeneity occurs and will show the sensitivity of the method.

3. Status

Just started.

Total cost:
  FF. 170,457
  u.a. 30,691
	E.C. Contribution: 46.92%
	  FF. 79,980
	  u.a. 14,400

COMMISSION OF THE EUROPEAN COMMUNITIES ENERGY R & D PROGRAMME **Objective:** Geothermal Energy	**Project or Sector:** Improvement of methods of exploration

Title:

    Application of linear programming for the determination of the Curie surface.

Duration: 18 months **Period:** 1/11/76 - 30/4/78	**Contract No:** 103-76 EGF **Project No:** G/B20(F)

Contractor: Institut National d'Astronomie et de Géophysique (C.N.R.S.)

Address: Place Eugène Bataillon,
          F 34060 Montpellier Cédex

Head of Project: Mr. Bayer (Université du Languedoc, Montpellier)

Description of research work

I. Objectives (aims)

To develop a new and reliable method for the calculation of the Curie surface on the basis of the geological, petrological and geophysical information available.

2. Work Programme

The contractor will develop a computer programme suitable for the introduction of this data. It will then be applied to the Massif Central where he will take into account all the magnetic data available. A comparison will be made between the results of this method with those of other authors.

3. Status

Just started.

Total cost:
   FF. 195,293
   u.a.   35,163
	E.C. Contribution: 51.2%
	   FF. 99,980
	   u.a. 18,001

COMMISSION OF THE EUROPEAN COMMUNITIES ENERGY R & D PROGRAMME **Objective:** Geothermal Energy	**Project or Sector:** Improvement of methods of exploration

Title:

    Study of the applicability of the geochemistry of gases in geothermal prospection.

Duration: 12 months **Period:** 31.12.76 - 30.12.77	**Contract No:** 112-76 EGI **Project No:** G/B35(I)

Contractor: Consiglio Nazionale delle Ricerche (C.N.R.)

Address: Lungarno Pacinotti, 55
        I 56100 Pisa

Head of Project: Dr. Franco D'Amore

Description of research work

I. Objectives (aims)

To identify the origin of the various gas components found in emanations, the possible relationships between them and the use that could be made of this analysis to determine reservoir temperatures.

2. Work Programme

The first stage will consist of a compilation of the data available for the field of Larderello, and a detailed analysis of it. Thermodynamic calculations will be made to formulate valid hypothesis on the origins of the various gases and the relationships between them. The study of the equilibria at various temperatures and pressures should make it possible to identify those reactions which could be used to estimate the temperature of reservoirs.

In the second stage, samples will be taken in the area of Siena and submitted to detailed analysis.

The conclusions of all temperature computations will be compared with the water geochemistry data and also with actual temperatures.

3. Status

Just started.

Total cost:
   Lit. 41,754,000
   u.a. 66,806
	E.C. Contribution: 50%
	   Lit. 20,877,000
	   u.a. 33,403

COMMISSION OF THE EUROPEAN COMMUNITIES ENERGY R & D PROGRAMME **Objective:** Geothermal Energy	**Project or Sector:** Improvement of methods of exploration

Title:

Application of trace element analysis to geothermal waters.

Duration: 12 months **Period:** 15/12/76 - 14/12/77	**Contract No:** 119-76 EGB **Project No:** G/B33(B)

Contractor: Universitaire Instelling Antwerpen

Address: Universiteipsplein, 1
2610 Wilrijk

Head of Project: Prof. R. Gijbels

Description of research work

I. Objectives (aims)

To use high-sensitivity trace element analysis of geothermal waters in order to obtain:

- additional information on mixing;

- indications on the possible nature of the reservoir rocks;

- to obtain additional information on the temperature in the reservoirs.

2. Work Programme

The following methods will be used:- mass spectrometry, radio chemistry after neutron analysis, X-ray fluorescence.

The water samples will come from various areas in France which are also under investigation by French organisations. The trace elements to be analysed will be selected on the basis of the analysis of major elements carried out by the partners in France.

The conclusions will indicate what additional interpretations could be provided by trace element analysis.

3. Status

Just started.

Total cost: FB. 4,010,000 u.a. 80,200	**E.C. Contribution:** 45% FB. 1,804,500 u.a. 36,090

COMMISSION OF THE EUROPEAN COMMUNITIES ENERGY R & D PROGRAMME **Objective:** Geothermal Energy	**Project or Sector:** Improvement of methods of exploration

Title:

Magnetic and magnetotelluric sounding in the Rhine Graben.

Duration: 12 months **Period:** 1/11/76 - 31/10/77	**Contract No:** 126-76 EGD **Project No:** G/B29(D)

Contractor: Universität Göttingen, Institut für Geophysik

Address: Herzberger Landstrasse 180, Postfach 876
D 34 Göttingen

Head of Project: Prof. U. Schmucker

Description of research work

I. Objectives (aims)

Utilisation of the possibilities of deep sounding by magnetic and magnetotelluric methods in order to establish if the electric anomaly in the Rhine Graben is caused by a bump in the earth's mantle.

2. Work Programme

The appropriate equipment, especially designed to cover periods (1,000 - 10 seconds), will be built and tested. The mathematical programmes for the analysis and interpretation of the data will be developed. Measurements will be made by both of these methods, along a profile perpendicular to the Rhine Graben at the latitude of Karlsruhe. The results of the measurements will be compared with information already available on this area.

3. Status

Just started.

Total cost: DM. 255,760 u.a. 69,880	**E. C. Contribution:** 43.6% DM. 111,511 u.a. 30,467

COMMISSION OF THE
EUROPEAN COMMUNITIES

ENERGY R & D PROGRAMME

Objective: Geothermal Energy

Project or Sector:

Improvement of methods of
exploration

Title:

Magnetotelluric and electric methods for geothermal
exploration.

Duration: 12 months

Period: 1/1/77 - 30/12/77

Contract No: 127-76 EGD

Project No: G/B30(D)

Contractor: Technische Universität Braunschweig

Address: Mendelssohnstrasse, 1
D 3300 Braunschweig

Head of Project: Prof. Dr. W. Kertz

Description of research work

I. Objectives (aims)

To investigate the applicability of the magnetotelluric and electric
sounding methods for the detection of temperature anomalies.

2. Work Programme

Equipment for magnetotelluric sounding will be designed and
constructed to cover a frequency range from 10 Hz to 10^{-5} Hz. The
equipment will include recorders and a computer for an initial
selection and treatment of the signals. Measurements with this
equipment will first be carried out in the area of Campi Flegrei with
the collaboration of scientists of the University of Naples. In addition
to the magnetotelluric measurements, electric sounding will be carried out
to determine the electric conductivity of the upper layers to a depth of
200 metres. On the basis of these measurements the temperature
distribution at great depth will be calculated with the help of a computer
programme. The calculated values will be compared with information
available from boreholes.

3. Status

Just started.

Total cost:	**E. C. Contribution:** 56%
DM 310,850	DM 169,942
u.a. 82,473	u.a. 46,432

COMMISSION OF THE EUROPEAN COMMUNITIES ENERGY R & D PROGRAMME Objective: Geothermal Energy	Project or Sector: Improvement of methods of exploration

Title:
Study of the origin and circulation of thermal waters in the Upper Rhine Graben by means of geochemical and isotopic methods.

Duration: 12 months	Contract No: 128-76 EGD
Period: 31.12.76 - 30.12.77	Project No: G/B37(D)

Contractor: Universität Tübingen

Address: Wilhelmstrasse 56,
D 7400 Tübingen

Head of Project: Prof. Friedrichsen

Description of research work

I. Objectives (aims)

To study the circulation of ground waters in the Rhine Graben.

2. Work Programme

Samples of water will be collected, at two weekly intervals, from the thirty boreholes which are available. These will be analysed for oxygen and hydrogen isotopes. Measurements of pH, CO_2 content and temperature will also be carried out. A few tritium and ^{14}C analyses will also be made. This data will be used to establish:

- the age of the waters in the boreholes;

- the circulation velocity of ground water;

- origin of the ground waters.

3. Status

Just started.

Total cost: DM. 270,880 u.a. 74,011	E.C. Contribution: 50% DM. 135,440 u.a. 37,005

COMMISSION OF THE EUROPEAN COMMUNITIES ENERGY R & D PROGRAMME **Objective:** Geothermal Energy	**Project or Sector:** Improvement of methods of exploration

Title:

Geothermal applications of the geochemical study of thermal springs in the Pyrenees.

Duration: 12 months **Period:** 1/1/77 - 31/12/77	**Contract No:** 147-76 ECF **Project No:** G/D22(F)

Contractor: Institut National d'Astronomie et de Géophysique (CNRS)

Address: Place Jussieu, 2 - Tour 53-54
               F 75221 Paris Cédex 05

Head of Project: Mr. Michard (University of Paris)

Description of research work

I. Objectives (aims)

To study the geochemistry of hot springs in the Eastern part of the Pyrenees and to determine the water temperature in the original reservoirs.

2. Work Programme

Analysis of major elements (which in this case includes sulphur compounds). Analysis of stable oxygen, hydrogen and carbon isotopes, as well as stable sulphur isotopes when appropriate.

Determination of the mixing schemes and calculation of the composition of waters coming from various depths. Calculation of equilibria between water and minerals in the reservoirs, and any modification during the rise to the surface. First estimate of temperatures at depth.

3. Status

Just started.

Total cost:
    FF. 475,200
    u.a. 85,560
	E.C. Contribution: 30.58%
	    FF. 145,350
	    u.a. 26,170

COMMISSION OF THE EUROPEAN COMMUNITIES ENERGY R & D PROGRAMME **Objective:** Geothermal Energy	**Project or Sector:** Improvement of methods of exploration

Title:

Microseismic exploration of geothermal anomalies.

Duration: 12 months **Period:** 31.3.77 - 30.3.78	**Contract No:** 177-77 ECD **Project No:** G/B12(D)

Contractor: Niedersächsisches Landesamt für Bodenforschung (NlfB Hannover)

Address: Alfred-Bentz-Haus, Postfach 51 01 53
D 3 Hannover 51

Head of Project: Dr. Steinwachs

Description of research work

I. Objectives (aims)

To investigate the applicability of the method for the detection of geothermal areas.

2. Work Programme

All the equipment necessary to carry out the measurements has to be constructed or purchased. It will be designed to be easily transportable. The method will have to be tested in areas where there is a known geothermal anomaly and the first one which is envisaged is the area of Urach.

3. Status

Just started

Total cost: DM. 325,000 u.a. 88,798	**E. C. Contribution:** 51,5% DM. 167,500 u.a. 45,765

COMMISSION OF THE EUROPEAN COMMUNITIES ENERGY R & D PROGRAMME **Objective:** Geothermal Energy	**Project or Sector:** Improvement of methods of exploration

Title:

Geochemical investigation of hot springs in Limagne (Massif Central), Vosges and the Eastern Pyrenees. Comparison of the geochemical methods.

Duration: 12 months	**Contract No:** 178-77 EGF
Period: 1.1.77 - 31.12.77	**Project No:** G/B38(F)

Contractor: Bureau de Recherches Géologiques et Minières (D.R.G.M.) (Paris)

Address: B.P. 6009
F 45018 Orleans Cédex

Head of Project: Mr. J. Halfon

Description of research work

I. Objectives (aims)

Geochemical study of waters of hot springs in the Limagne (Clermont Ferrand), Eastern Pyrenees and Vosges areas.

2. Work Programme

Analysis will be carried out in the following order of priority:

major elements, trace elements, dissolved gases, stable isotopes, radioactive isotopes.

From these analyses the contractor will calculate:-

- mixing with waters from the surface;

- temperature in the reservoir;

- age of waters;

- nature and volume of reservoir.

The nature of the information which can be deduced from the various analyses will be discussed in detail, in the case of the hot springs studied in this project and in other projects executed by teams in France.

3. Status
Just started.

Total cost:	**E.C. Contribution:** 50%
FF. 870,468	FF. 435,234
u.a. 156,728	u.a. 78,364

COMMISSION OF THE EUROPEAN COMMUNITIES ENERGY R & D PROGRAMME Objective: Geothermal Energy	Project or Sector: Improvement of methods of exploration

Title:

Application of the system Rb-Sr for the study of geothermal springs.

Duration: 12 months Period: 4.4.77 - 3.4.78	Contract No: 180-77 EGF Project No: G/B25(F)

Contractor: Institut National d'Astronomie et de Géophysique

Address: Place Jussieu, 4 - Tour 14
F 75230 Paris Cédex 05

Head of Project: Prof. C. J. Allègre (Université de Paris VI)

Description of research work

I. Objectives (aims)

To determine the relationship between the isotopic ratio $^{87}Sr/^{86}Sr$ in thermal waters and in the rocks in which they are contained.

2. Work Programme

Water samples will be collected in the Massif Central from both cold springs and hot springs whose origins are best defined. Isotop analysis will be carried out for Sr, Rb and K. Typical rocks from the Massif Central will also be analysed (volcanic rocks from the Cantal, granites and crystalline schists, sedimentary rocks). The distribution of Sr, Rb and K between water and rocks will be calculated for various conditions of equilibrium.

3. Status

Just started.

Total cost: FF. 239,574 u.a. 43,133	E.C. Contribution: 32.4% FF. 77,610 u.a. 13,973

COMMISSION OF THE EUROPEAN COMMUNITIES ENERGY R & D PROGRAMME **Objective:** Geothermal Energy	**Project or Sector:** Improvement of methods of exploration

Title:

Reconstruction of the trace element distribution in the late Quartenary differentiated magma chamber of the Laacher volcano (East Eifel).

Duration: 24 months **Period:** 1/1/77 - 30/12/78	**Contract No:** 219-77 EGB **Project No:** G/E39(B)

Contractor: Universitaire Instelling Antwerpen

Address: Universiteitsplein, 1
2610 Wilrijk

Head of Project: Prof. R. Gijbels

Description of research work

I. Objectives (aims)

To contribute to the understanding of the differentiation mechanism which occurs in a magma chamber. The work will be carried out in collaboration with the University of Bochum.

2. Work Programme

The analysis will be carried out by neutron activation and gamma counting with the help of a low energy detector. Whole rock samples and isolated minerals will be analysed. Initially, a large spectrum of elements will be investigated. It will then be restricted to those which appear to be most useful for the evaluation of partitioning coefficients during partial crystallization.

3. Status

Just started.

Total cost:
  FB 1.200.000,--
  24.000 u.c.
	E.C. Contribution:
	  FB 560.040,--
	  11.200 u.c.

COMMISSION OF THE EUROPEAN COMMUNITIES ENERGY R & D PROGRAMME Objective: Geothermal Energy	Project or Sector: Improvement of methods of exploration

Title: Research and characterisation of seismic noise in geothermal areas.	

Duration: 12 months Period:	Contract No: 220-77 EGF Project No: G/B21(F)

Contractor: Institut National d'Astronomie et de Géophysique (INAG)

Address: 77, Avenue Denfert-Rochereau, 75014 Paris

Head of Project: Mr. M. A. Choudhury (Institut de Physique du Globe, Strasbourg)

Description of research work

  I.   Objectives (aims)

  The objective of this project is the registering of seismic noises and microearthquakes in geothermal areas, in order to determine the possible existance of special seismic activity in these areas.

  2. Work Programme

  The seismic noise in the 1-500 Hz range will be registered on magnetic tape using three mobile measuring units.   The influence of night and day as well as seasonal influences will be taken into consideration. The noise will be studied in two different geothermal regions to see if there are regional differences.

  3. Status

  Being signed.

Total cost:
     FF.   176,420
     u.a.   31,764
	E.C. Contribution:
	     FF. 73,800
	     u.a. 13,287

Project C

Sources of hot water (low enthalpy)

COMMISSION OF THE EUROPEAN COMMUNITIES ENERGY R & D PROGRAMME **Objective:** Geothermal Energy	**Project or Sector:** Sources of hot water (low enthalpy)

Title:

Investigation of the optimal use of geothermal waters for the heating of several types of dwelling in various European climates.

Duration: 12 months **Period:** 1/1/77 - 30/12/77	**Contract No:** 077-76 EGF **Project No:** G/C14(F) and G/C16(F)

Contractor: Electricité de France (EDF), et L'ingénerie Française du Tout Electrique (INFRATEL)

Address: EDF - Quai Watier, 6 - Boîte Postale no 24, F 78400 Chatou
INFRATEL - Allés de Clichy, 21 - F 93340, Le Raincy

Head of Project: Mr. Aurcille (EDF), Mr. Villaume (INFRATEL)

Description of research work

I. Objectives (aims)

The study is based on previous work undertaken on the geothermal heating of apartments under the conditions specific to the area of Paris. The present project should consider a broader range of possibilities.

2. Work Programme

The following parameters will be investigated:-

- climate;

- type of apartment (nature of insulation and ventilation);

- various heating systems;

- extent of individual regulation;

- design of the central heating plant (heat exchanges, heat pumps);

- booster boiler;

- characteristics of the geothermal source;

- re-injection;

- cost of the heating systems in the apartments;

../..

Total cost:	**E.C. Contribution:** 50%
FF 535,000	FF 267,500
u.a. 96,327	u.a. 48,164

COMMISSION OF THE EUROPEAN COMMUNITIES ENERGY R & D PROGRAMME **Objective:** Geothermal Energy	**Project or Sector:** Sources of hot water (low enthalpy)

- investments (plant, wells).

3. Status

Just started.

COMMISSION OF THE EUROPEAN COMMUNITIES ENERGY R & D PROGRAMME Objective: Geothermal Energy	Project or Sector: Sources of hot water (low enthalpy)

Title:

Well tests in water-dominated geothermal systems - interpretation.

Duration: 12 months Period: 1/1/77 - 31/12/77	Contract No: C06-76 EGI Project No: G/01(I)

Contractor: Ente Nazionale per l'Energia Elettrica "ENEL" Roma

Address: Piazza Bartolo da Sassoferrato, 14
I 56100 Pisa

Head of Project: Ing. A. Barelli (Centro di Ricerca Geotermica, Pisa)

Description of research work

I. Objectives (aims)

Experimental study of the thermodynamics and hydrodynamics of a two-phase flow in a geothermal well.

2. Work Programme

The work will be done on well ALFINA 7. Equipment will be installed at the well head to separate the liquid and the gaseous phase and to measure all useful parameters. Pressure and water levels will be measured in five neighbouring wells in order to follow the effect of flow conditions in well ALFINA 7.

The results of the experiment will be compared with existing models which will eventually be improved when necessary.

3. Status

Just started.

Total cost: Lit. 104,900,000 u.a. 167,840	E.C. Contribution: 50% Lit. 52,450,000 u.a. 83,920

| COMMISSION OF THE EUROPEAN COMMUNITIES

ENERGY R & D PROGRAMME

Objective: Geothermal Energy | **Project or Sector:**

Sources of hot water
(low enthalpy) |

Title:

Study of the chemical reactions occuring in the course of the exploitation of a geothermal doublet. Application to the warm waters of the Dogger of the Paris Basin.

| **Duration:** 12 months
Period: 1/1/77 – 30/12/77 | **Contract No:** 092-76 EGF
Project No: G/010(F) |

Contractor: Bureau des Recherches Géologiques et Minières (B.R.G.M.)

Address: B.P. 6009
F 45018 Orléans Cédex

Head of Project: Mr. Leleu (Service Géologique National, Orléans)

Description of research work

I. Objectives (aims)

To study the chemical reactions of waters and rocks in the reservoir and the deposits from geothermal water during its cooling in heat exchanges.

2. Work Programme

(a) study of the chemistry of the warm water and of the geothermal reservoir;

(b) study of the equilibria on the basis of information available from literature;

(c) calculation of the possible encrustations in heat exchanges and other circuits, of reaction with reservoir rock and of the re-injected water;

(d) review of the information available on the kinetics of these reactions;

(e) construction of two experimental pieces of equipment for kinetic measurements.

3. Status

Just started.

| **Total cost:**
  FF 334,000
  u.a. 60,136 | **E.C. Contribution:** 50%
  FF 167,000
  u.a. 30,068 |

COMMISSION OF THE EUROPEAN COMMUNITIES ENERGY R & D PROGRAMME **Objective:** Geothermal Energy	**Project or Sector:** Sources of hot water (low enthalpy)

Title:

Study of the influence of physical characteristics of the aquifer and of the neighbouring rocks on the water temperature in the production well of a geothermal doublet.

Duration: 12 months **Period:** 15/12/76 - 14/12/77	**Contract No:** 093-76 EGF **Project No:** G/C7(F)

Contractor: Bureau de Recherches Géologiques et Minières (B.R.G.M.)

Address: B.P. 6009
F 45018 Orléans Cédex

Head of Project: Mr. A. Gringarten (Service Géologique National, Orleans)

Description of research work

I. Objectives (aims)

To develop improved methods of calculation for reservoir engineering.

2. Work Programme

The following parameters will be studied:

- thermal conductivity in horizontal and vertical directions;

- thermal conductivity of the neighbouring rocks;

- variation of the density with temperature;

- variation of viscosity with temperature.

The influence of each parameter will be studied in detail, and in particular on the date at which the temperature at the production well will start decreasing.

3. Status

Just started.

Total cost: FF 299,000 u.a. 53,835	**E.C. Contribution:** 50% FF 149,500 u.a. 26,918

COMMISSION OF THE EUROPEAN COMMUNITIES ENERGY R & D PROGRAMME **Objective:** Geothermal Energy	**Project or Sector:** Sources of hot water (low enthalpy)

Title:

Study of the performance of the geothermal heating installation at Creil. Thermal and economic balance during the first period of heating.

Duration: 9 months **Period:** 14/12/76 - 13/9/77	**Contract No:** 094-76 EGF **Project No:** G/C8(F)

Contractor: Electricité de France (EDF)

Address: Quai Watier, 6 - Boîte Postale N° 24
F 78400 Chatou

Head of Project: Mr. Aureille (Service Ensembles de Production)

Description of research work

I. Objectives (aims)

To study in detail the performance of the central heating installation during its first season of operation.

2. Work Programme

(a) design and installation of measurement instruments:-

flowmeters and thermocouples are necessary to determine the thermal power supplied by a geothermal well, the boilers and the heat pumps on the one hand, and the power required by the group of dwellings on the other. A data acquisition system will also be set up;

(b) measurements and calculations:-

particular attention will be paid to fluctuations during day and night, variations over a week and the influence of climatic conditions. The energy balance will be calculated periodically.

3. Status

Just started.

Total cost: FF. 159,000 u.a. 28,628	**E.C. Contribution:** 50% FF. 79,500 u.a. 14,314

COMMISSION OF THE EUROPEAN COMMUNITIES ENERGY R & D PROGRAMME **Objective:** Geothermal Energy	**Project or Sector:** Sources of hot water (low enthalpy)

Title:

Study on a mathematical model of the influence of heterogeneities of a geothermal deposit on its behaviour during exploitation.

Duration: 12 months **Period:** 1/1/77 - 31/12/77	**Contract No:** 099-76 EGF **Project No:** G/C6(F)

Contractor: Association pour la Recherche et le Développement des méthodes et processus industriels (ARMINES)

Address: rue Saint-Honoré, 35
                  F 77305   Fontainebleau

Head of Project: Mr. G. de Marsily (Laboratoire d'Hydrogéologie
                                          Mathématique)

Description of research work

I. Objectives (aims)

To study the influence on a doublet of heterogeneities in the stratified layers of a reservoir.

2. Work Programme

The variations of the stratified layers will be characterised by permeability, conductivity, number and thickness of impermeable layers within the permeable aquifer.   The results will be compared with the behaviour of a homogeneous aquifer.

3. Status

Just started.

Total cost:
   FF.   214,400
   u.a.   38,603
	E.C. Contribution: 75%
	   FF. 160,800
	   u.a. 28,952

Project D

Steam sources (high enthalpy) and hot rocks

COMMISSION OF THE EUROPEAN COMMUNITIES ENERGY R & D PROGRAMME Objective: Geothermal Energy	Project or Sector: Steam sources (high enthalpy)

Title:

        Methodology and interpretation of well production tests
        in the case of two-phase flow.

Duration: 12 months **Period:** 11/11/76 - 10/11/77	**Contract No:** 081-76 EGF **Project No:** G/D15(F)

Contractor: Bureau de Recherches Géologiques et Minières (ORLEANS)

Address: B.P. 6009
            F 45018 Orleans Cédex

Head of Project: Mr. A. Gringarten (B.R.G.M. Service Géologique National)

Description of research work

I. Objectives (aims)

To develop a mathematical model to describe the hydrodynamic behaviour
of high pressure water which is partially vaporised in the course of
its flow to the surface.

2. Work Programme

All data available from well tests carried out in various geothermal
fields in the world will be compiled.   The model will be checked by
means of this data.   An attempt will be made to determine the
characteristics of the reservoir from the measurements at the well head.

3. Status

Just started.

Total cost:
   FF.   115,800
   u.a.   20,850
	E.C. Contribution: 50%
	   FF. 57,900
	   u.a. 10,425

COMMISSION OF THE EUROPEAN COMMUNITIES ENERGY R & D PROGRAMME Objective: Geothermal Energy	Project or Sector: Steam sources (high enthalpy)

Title:

    Research on drilling muds and on cement mixtures to be used in geothermal wells at high temperatures.

Duration: 12 months Period: 15/12/76 - 14/12/77	Contract No: 089-76 EGI Project No: G/D9(I)

Contractor: AGIP S.p.A. Attivita Minerarie

Address: C.P. 4174
        I 20100 Milano

Head of Project: Dr. F. Monardi

Description of research work

I. Objectives (aims)

To study on a laboratory scale the properties of improved muds and improved cements to be used in geothermal wells at temperatures above 200°C.

2. Work Programme

(a) Drilling muds:- existing information will be compiled from literature.

Improvements on the most advanced compositions known and used at the present time will be tested on a laboratory scale. Both aqueous and oil-based muds will be investigated.

Tests will be carried out at temperatures above 140°C and pressures from 100 - 500 kg/cm^2. The viscosity, density and stability will be studied on muds before and after the aging process. The effect of vibrations will also be studied.

(b) Cements:- data will be collected from literature on cements for high temperature wells. Particular attention will be given to the special problems which occur with the use and behaviour of these cements in geothermal applications.

The experimental work will consist of studying new compositons for temperatures up to 300°C. The behaviour of the mortars in the various stages of handling, injection in the well, setting and hardening will be considered. ../..

Total cost:
   Lit. 58,000,000
   u.a. 92,800
	E.C. Contribution: 50%
	   Lit. 29,000,000
	   u.a. 46,400

COMMISSION OF THE EUROPEAN COMMUNITIES ENERGY R & D PROGRAMME **Objective:** Geothermal Energy	**Project or Sector:** Steam sources (high enthalpy)

The mechanical properties of the fully hardened mortars, and their change with age, will be evaluated under conditions similar to those existing in a well.

3. Status

Just started.

COMMISSION OF THE EUROPEAN COMMUNITIES ENERGY R & D PROGRAMME **Objective:** Geothermal Energy	**Project or Sector:** Hot dry rocks

Title:

    Study of acoustic and microseismic emissions associated with a hydraulic fracture.

Duration: 12 months **Period:** 1/1/77 - 31/12/77	**Contract No:** 100-76 EGF **Project No:** G/D14(F)

Contractor: Institut National d'Astronomie et de Géophysique (CNRS)

Address: Place Jussieu, 4 - Tour 14
    F 75230 Paris Cédex 05

Head of Project: Prof. G. Jobert (University of Paris)

Description of research work

I. Objectives (aims)

To investigate the feasibility of determining the position and extension of a fracture by means of geophones at the surface.

2. Work Programme

A granite body will be selected in which a borehole already exists to a depth of 200 metres. The signals will be detected at the surface by means of:

- geophones optimised for a frequency of 5 - 60 Hz;

- hydrophones for the range of 50 Hz to 50 kHz.

The signals will be interpreted by means of computer codes which are already in existence.

3. Status

Just started.

Total cost:
  FF. 258,000
  u.a. 46,453
	E.C. Contribution: 58.14%
	  FF. 150,000
	  u.a. 27,008

COMMISSION OF THE EUROPEAN COMMUNITIES ENERGY R & D PROGRAMME Objective: Geothermal Energy	Project or Sector: Hot dry rocks

Title:

    Study of the possibilities of stimulating permeability in crystalline rock by chemical leaching.

Duration: 24 months Period: 15/12/76 - 14/12/78	Contract No: 164-76 EGF Project No: G/D12(F)

Contractor: Institut Français du Pétrole

Address: 1 et 4 av. de Bois-Préau - B.P. 311
        F 92506 Rueil-Malmaison

Head of Project: Mr. P. Simandoux

Description of research work

I. Objectives (aims)

To investigate whether it is possible to increase the permeability, by chemical leaching, of a crystalline rock after hydraulic fracturation. This project should contribute to the feasibility of the hot dry rock concept.

2. Work Programme

Equipment will be constructed to test the behaviour of rock samples in different media under carefully controlled conditions. In particular, fluid will be circulated at controlled rates, at high pressures, and temperatures between 150 and 220°C.

Three or four types of rock will be tested, and in particular granites and gneisses. Sodium carbonate and hydrofluoric acid will be selected initially for the chemical leaching. Other media will eventually be selected on the basis of the first results.

3. Status

Just started.

Total cost:
  FF. 937,496
  u.a. 168,797
	E.C. Contribution: 50%
	  FF. 468,748
	  u.a. 84,398

COMMISSION OF THE EUROPEAN COMMUNITIES ENERGY R & D PROGRAMME Objective: Geothermal Energy	Project or Sector: Steam sources (high enthalpy)

Title:

Improvement of equipment for measurements in geothermal wells.

Duration: 12 months Period: 31.12.76 - 30.12.77	Contract No: 166-76 EGI Project No: G/D4(I)

Contractor: Ente Nazionale per l'Energia Elettrica (ENEL)

Address: Piazza Bartolo da Sassoferrato, 14
I 56100 Pisa

Head of Project: Dr. Ferrara (Direzione degli Studi e Ricerche, Centro di Ricerca Geotermica)

Description of research work

I. Objectives (aims)

To develop a complete system with pressure and temperature probes to carry out measurements at various levels of a well and with recording equipment at the surface. The probes and the cable should be able to operate satisfactorily in steam at temperatures up to 300°C and pressures up to 200 bars.

2. Work Programme

(a) temperature and pressure probes will be constructed and tested in autoclaves. Their reliability and the accuracy of the measurements will be checked;

(b) a cable with seven conductors will be selected and its behaviour at the specified levels of temperature and pressure will be investigated;

(c) the reliability of the seal between cable and probe will be checked. A leak-proof stuffing box will be developed to allow movement of the cable in and out of the well head;

(d) each element will be tested individually, and eventually the complete equipment will be built and tested in an actual well.

3. Status

Just started.

Total cost: Lit. 96,280,000 u.a. 154,048	E.C. Contribution: 50% Lit. 48,140,000 u.a. 77,024

COMMISSION OF THE EUROPEAN COMMUNITIES ENERGY R & D PROGRAMME **Objective:** Geothermal Energy	**Project or Sector:** Steam sources (high enthalpy)

Title:

Rock fracturation tests at great depth by means of hydraulic impulse.

Duration: 12 months **Period:** 31.12.76 - 30.12.77	**Contract No:** 167-76 EGI **Project No:** G/D6(I)

Contractor: Ente Nazionale per l'Energia Elettrica (ENEL)

Address: Piazza Bartolo da Sassoferrato, 14
I 56100   Pisa

Head of Project: Dr. Barelli (Direzione degli Studi e Ricerche / Centro di Ricerca Geotermica)

Description of research work

I. Objectives (aims)

To study the feasibility of increasing the permeability of deep-seated formations of sedimentary rock by cyclic variations of hydraulic pressure.

2. Work Programme

The work will be carried out in a well in the Travale area at a depth of 1800 metres.   The rock types are limestone and phyllade.   The well will first be submitted to a constant pressure of 20 kg/cm^2 for about twenty days in order to observe any variation in permeability. Fracturation tests will be carried out with additional pressure of at least 10 kg/cm^2 at periodicities of a maximum of 10 minutes.   Each test, under uniform conditions, will last from 20 - 40 days.   After each testing period the permeability will be measured.   In the case of success, the economic feasibility of the process will be compared with the normal fracturation method.

3. Status

Just started

Total cost:
    Lit.   60,060,000
    u.a.   96,096
	E.C. Contribution: 50%
	    Lit. 30,030,000
	    u.a. 48,048

COMMISSION OF THE EUROPEAN COMMUNITIES ENERGY R & D PROGRAMME **Objective:** Geothermal Energy	**Project or Sector:** Steam sources (high enthalpy)

Title:

Study of the geodynamic effects of hydraulic impulses in deep wells.

Duration: 12 months **Period:**	**Contract No:** 203-77 EGI **Project No:** G/D1(I)

Contractor: Università degli Studi di Napoli

Address: Largo S. Marcellino, 10
I 80138 Napoli

Head of Project: Prof. Oliveri del Castillo

Description of research work

I. Objectives (aims)

The project will be coupled with the fracturation tests made on the Travale well by ENEL. It is intended to measure the effects of the fracturation at the surface by topographic measurements. In addition, extensive seismic observations will be made in order to detect the horizontal extension of the fractured area.

2. Work Programme

A network of reference points will be established around the area to be fractured and a system for automatic levelling will be installed. The water level difference will be recorded continuously during fracturation and will be recalibrated periodically.

Seismic measurements will be made with high sensitivity geophones. Six geophones will be installed in permanent stations. Four others will be mobile. All indications will be transmitted by radio to a central station equipped for recording and for the initial treatment of the signals. Before fracturation, the seismicity of the area will be checked in order to establish the noise level.

3. Status
Being signed.

Total cost: Lit. 65,657,000 u.a. 105,051	**E.C. Contribution:** 50% Lit. 32,828,500 u.a. 52,525

COMMISSION OF THE EUROPEAN COMMUNITIES ENERGY R & D PROGRAMME Objective: Geothermal Energy	Project or Sector: Steam sources (high enthalpy)

| Title:

Development of computer methods to study the inducement of cracking in hot rocks by thermal or hydro-pressure means. ||

Duration: 12 months Period:	Contract No: 218-77 EG UK Project No: G/D16(UK)

Contractor: Central Electricity Generating Board (C.E.C.B.)

Address: GD Berkeley, Gloucestershire GL 13 9 PB

Head of Project: D. J. Lewis

Description of research work

I. Objectives (aims)

To adapt, for the study of rocks, an existing computer code called CEGB BERSAFE, which is normally used for the study of the propagation of cracks in metals.

2. Work Programme

Data will be collected from literature on the fracture behaviour of rocks typical of geothermal situations, including granite. The code will be modified to describe the rock fracture propagation. In the case of success, the code will be applied to the fracturation of a deep-seated rock body by hydraulic pressure applied in the borehole. Two cases will be studied:

- conditions typical of oil exploitation;

- a crystalline rock body, in particular granite.

The calculations will provide the pressure required to propagate the crack and the extension of the crack as a function of pressure. It is also hoped to evaluate the extent of thermal cracking at right-angles to the main crack in the case of water circulation.

3. Status

Being signed.

Total cost: £ 13,500 u.a. 32,400	E.C. Contribution: 50% £ 6,750 u.a. 16,200

PROPOSALS
UNDER NEGOTIATION

PROPOSALS UNDER NEGOTIATION

Project A: Acquisition and collation of existing and new geothermal data

1. Correlation between the chemical composition of the surface deposits and the condition of circulating subterranian fluids in the volcanic area of Lazio.

   Università degli Studi di Roma, Italy

Project B: Improvement of methods of exploration

1. Study and realisation of a measuring probe to determine directly the geothermal flux in situ.

   Institut National d'Astronomie et de Géophysique, France

Project C: Sources of hot water (low enthalpy)

1. Feasibility study and demonstration of the use of a thermo-hydraulic loop for the conversion of geothermal energy (low enthalpy) into electricity.

   Ente Nazionale per l'Energie Elettrica ENEL, Italy

Systems analysis: development of models

Since the energy price crisis many decisions and choices to be made
at various political levels are strongly influenced by energy
considerations.   Some of these decisions relate to the short-term
tactical aspect, while others concern the long-term strategy.   All of
them are governed by a great number of constraints, resulting from the
many factors to be taken into consideration, the complexity of their
interrelationships and the way they vary with time.

The principal aim of the EC Systems Analysis programme is to develop
a tool which shall serve the needs of the Community in identifying and
analysing problems in the field of energy and related sectors (investigation of alternative energy strategies, guidance for energy R & D).
This will be of use for the Governments of the Member States as well
as the Commission.

As the Member States' economies are closely interlinked, tackling of
energy problems on a national level is no more sufficient.   It
appeared, however, appropriate to build the European modelling
exercise on certain existing national models which can be integrated
to a Community model.   It must be remembered that the model(s) to be
developed are not intended to become decision-making tools.   They
should mainly provide decisions with comprehensive and coherent
quantitative information based on explicit hypotheses.

The first six months of the programme were mostly devoted to define
in sufficient detail the practical research work to be undertaken.
After this, the Commission, assisted by the ACPM and a working group
of specialists, decided to start work on four different levels but to
put its main emphasis on the development of a medium-term quasi-
dynamic energy model in which different economy-related submodels will
be coupled with the so-called "energy flow optimization model".
The chosen structure should enable its users to simulate the reactions
of energetic systems and the economic subsystem (of the EC as a whole,
as well as of the individual member countries) in response to
quantitative and/or qualitative changes in the energy supply system.
On the other hand, this structure also constitutes a tool for assessing
the ability of the energy supply system to react upon changes
(constraints) caused by economical, technological, ecological or
resource availability events.

This whole work has just started and will in a first round be confined
to France as a pilot country.

The contracts concluded, therefore, are essentially covering the
following subjects:

1.   the development and improvement of the above mentioned energy
     flow optimization model;

2. the adaptation of a macro-economic model computing final demand
    and total energy consumption for given reference years;

3. the updating and adaptation of an input/output model describing
    the inter-industry relationships required for coherent short
    and medium term projections of intermediate and final demands
    of energy;

4. the setting up of a comprehensive European data base containing
    the data necessary for describing the structures and components
    of the system.

Besides this work, which will be completed towards the end of the
second phase (and after having concluded quite some additional
contracts with other institutes, in order to extend the model to
all EC countries) further contracts have already been arranged
concerning:

5. an integrated dynamic simulation model for studying at medium
    or long term the effects of new energy technologies or of
    certain policy measures taken at EC level;

6. the study of the effects on the world energy trade system of
    certain environmental or conservational policy decisions;

7. investigation of certain methodological problems related to
    energy system's "modelling" in general.

As already stated, this programme is based on a certain number of
choices made after thorough reflection and discussion.   Although
the first test-runs for the energy flow optimization model are
expected to take place (for the pilot country) still in 1977, some
very important additional contributions are required in phase II
before the model can be used for the purposes for which it has
been conceived.

SUMMARY AND BREAKDOWN

OF FUNDING

Objective: Systems Analysis: Development of Models

	Number of contracts(*)	Total cost u.a.	E.C. contribution u.a.	Number of proposals under negotiation
	10	652,424	602,424	-
Total	10	652,424	602,424	-

(*) Signed both by the Commission and the Contractor or sent for signature to the Contractor.

COMMISSION OF THE EUROPEAN COMMUNITIES ENERGY R & D PROGRAMME Objective: Systems Analysis: Development of models	Project or Sector:

Title: Reference Energy System Data Base and its Integration with the Energy Flow Optimization Model (Phase I).	

Duration: 12 months Period: 1/1/76 - 31/12/76	Contract No: 010-76 EMB Project No:

Contractor: Systems Europe (SE) Address: Avenue de Broqueville, 194 - 1200 Brussels, Belgium Head of Project: E. Jamoulle	

Description of research work

I. Objectives (aims)

The so-called "Level II Modeling" was defined and recommended in the "Report of the ad hoc Expert Group to ACPM on Systems Analysis and Modeling in the Field of Energy", dated 13 November, 1975. The main components of the envisaged Level II Model were outlined in the "Final Report; SE Doc. 319.31", dated 19 December, 1975, delivered to EEC, along with an overall four phase plan for its realization. Three main model components were distinguished, as follows:

A. The Reference Energy System (RES) and Data Base

B. The Energy Flow Optimization Model (EFOM)

C. The I/O Model

The objective of the project is to furnish EEC with some in-house Level II Modeling capability by the end of Phase I, to allow some preliminary energy policy analyses, and to make any necessary modifications to the remaining phases of the Level II Modeling programme based on the results of Phase I and other EEC energy modeling activities.

../..

Total cost: FB. 6,500,000 u.a. 130,000	E. C. Contribution: 100% FB. 6,500,000 u.a. 130,000

COMMISSION OF THE EUROPEAN COMMUNITIES ENERGY R & D PROGRAMME Objective: Systems Analysis: Development of Models	Project or Sector:

2. Work Programme

A. RES – Data Base Tasks

    Task A1: Data Base Planning

    Task A2: Prototype Data Base Design, DB Management Software and Simulation Software Design

    Task A3: Data Collection for France

    Task A4: Case Studies

B. EFOM Tasks

    Cooperation with IEJE particularly for RES-DB and EFOM interfaces

C. I/O Tasks

    Cooperation with Battelle particularly for RES-DB and EXPLOR interfaces

3. Status

Completed.

COMMISSION OF THE EUROPEAN COMMUNITIES ENERGY R & D PROGRAMME Objective: Systems analysis: development of models	Project or Sector:

Title:

Utilization of large scale mathematical programming techniques in energy optimization problems for the Community.

Duration: 6 months Period: 1/5/76 - 31/10/76	Contract No: 011-76 EMB Project No:

Contractor: Université Catholique de Louvain (U.C.L.)

Address: 1, Place de l'Université, Louvain La Neuve, Belgium

Head of Project: M. Smeers

Description of research work

I. Objectives (aims)

Utilization of large scale mathematical programming techniques in energy optimization problems for the Community. The study considers two methods: nested decomposition (a) (b) and block factorization (c). These methods have been tested by means of experimental codes on several problems presenting a structure similar to the Community energy optimization problems.

References:

(a) Ho, J. (1974): "Nested decomposition of large scale linear programmes with the staircase structure", Technical Report 74-4, Systems Optimization Laboratory, Department of Operation Research, Stanford University.

(b) Loute, E. (1976): "A revised simplex method for block structured linear programmes", thesis, Faculté des Sciences Appliquées, University Catholique de Louvain.

(c) Tomlin, J.A.: "Notes on a Workshop on Energy Systems Modelling", Session 3 (February 10-11, 1975), Systems Optimization Laboratory, Department of Operation Research, Stanford University.

../..

Total cost: FB. 750,000 u.a. 15,000	E.C. Contribution: 100% FB. 750,000 u.a. 15,000

COMMISSION OF THE EUROPEAN COMMUNITIES ENERGY R & D PROGRAMME Objective: Systems Analysis: development of models	Project or Sector:

2. Work Programme

A preliminary report has been completed: 'Utilization of mathematical programming techniques for large scale problems applied to energy optimization', in January 1977.

More complete numerical experiences have been carried out (with particular reference to calculation strategies). Some experiments using the procedures of MPSX code from IBM have been done, in order to implement the use of the method with a standard code.

3. Status

Numerical computations are achieved and the final report is currently being written.

COMMISSION OF THE EUROPEAN COMMUNITIES ENERGY R & D PROGRAMME Objective: Systems analysis: development of models	Project or Sector:

Title: Study of energy policy measures.	

Duration: 6 months Period: 1/5/76 - 31/10/76	Contract No: 012-76 EMB Project No:

Contractor: Université Libre de Bruxelles (ULB) Address: Ave. F.D. Roosevelt, 50 - 1050 Bruxelles, Belgium Head of Project: Mrs. Thys	

Description of research work

I. Objectives (aims)

This is a feasibility study consisting of the theoretical analysis of the introduction of policy measures into a macroeconomic model that can be used for both optimization and simulation purposes.

The study is based on a European econometric model that is already operational[1].   This model has been modified to take account of:
(a) energy variables as regards both supply (imports, prices) and final demand;

(b) a link with the EXPLOR input-output model.

---

[1] A. Dramais: "Neuf modèles nationaux liés pour l'étude de la diffusion des fluctuations conjoncturelles et des effets des mesures de politique économique entre les pays membres du Marché Commun", Cahiers Economiques de Bruxelles Nos. 64-65-66, 1974 and 1975.

../..

Total cost:
    BF.   750,000
    u.a.   15,000
	E. C. Contribution: 100%
	    BF. 750,000
	    u.a. 15,000

COMMISSION OF THE EUROPEAN COMMUNITIES ENERGY R & D PROGRAMME Objective: Systems analysis: development of models	Project or Sector:

I. Objectives (aims) cont/..

A methodological analysis has been carried out to ensure compatibility between the macroeconomic approach of the national account type and the input-output analysis.

2. Work Programme

Two reports have been submitted:

The first is: "Adaptation du modèle Desmos aux problèmes énergétiques" by A. Dramais and F. Thys-Clément, November 1976.

The second is: "Eureca; un modèle d'équilibre général mesurant l'impact des variables énergétiques sur l'environnement économique", by A. Dramais and F. Thys-Clément, January 1977.

The second report deals in general with the problems of the link between macroeconomic models (national accounts type) and the input-output approach.

3. Status

The contract is completed.

COMMISSION OF THE EUROPEAN COMMUNITIES ENERGY R & D PROGRAMME Objective: Systems analysis: development of models	Project or Sector:

Title:

Improvement of the energy flow optimization model.

Duration: 12 months Period: 1/1/76 - 31/12/76	Contract No: 036—76 EMF Project No:

Contractor: Institut Economique et Juridique de l'Energie

Address: Domaine Universitaire de Grenoble — Saint Martin d'Hères, France

Head of Project: Mr. J.M. Martin

Description of research work

I. Objectives (aims)

Definition of an energy system optimization model (MOSE) for a given country based on the improvement of the ENERGIE model constructed and used by the IEJE between 1970 and 1975.   This improvement relates both to the representation and formalization of the various energy chains and to the modification of the software used (SCORPION and OSIRIS).   As regards representation, the aim is to include new energy processes and vectors and to extend the model to industries that are heavy consumers of energy.

It is also necessary to carry out some case studies in order to validate the model and study its sensitivity.   It has to be connected to the data base established by Systems Europe and to the EXPLOR macroeconomic model.

2. Work Programme

(a) Methodological study of multiperiod optimization (as compared with snapshot optimization), multinational optimization and use of the dual;

(b) Introduction of new vectors (hot water, hydrogen, steam) and new processes (nuclear, hydrogen production, solar heating, geothermal heating, district heating, etc.);

(c) Extension of the optimization to high energy—consuming processes in industry;

../..

Total cost:
 	
   FF.   499,850
   u.a.   89,998
	E.C. Contribution: 100%
	
	   FF. 499,850
	   u.a. 89,998

COMMISSION OF THE EUROPEAN COMMUNITIES ENERGY R & D PROGRAMME Objective: Systems analysis: development of models	Project or Sector:

2. **Work Programme** cont/..

(d) Collection of data on the French case;

(e) Participation in data specification;

(f) Specification of the mathematical formalization of the PL;'

(g) Improvement of the SCORPION input and output software, use of OSIRIS to output tables, writing of the interface with the data base;

(h) Completion of three case studies and analysis of the results.

3. **Status**

The various tasks have been completed and are the subject of the final report.

COMMISSION OF THE EUROPEAN COMMUNITIES ENERGY R & D PROGRAMME Objective: Systems analysis: development of models	Project or Sector:

| Title:

    Implementation of a dynamic energy model for the European Community. ||

Duration: 12 months Period: 12/10/76 - 11/10/77	Contract No: 048-76 EMD Project No:

Contractor: Kernforschungsanlage Jülich

Address: D-517 Jülich 1, Postfach 1913, Germany

Head of Project: M. U. Schoeler

Description of research work

I. Objectives (aims)

The objective of the project is to develop an energy model which can give a decision aid for defining a reasonable energy policy. With the help of the model one can simulate strategies:

- for developing the energy supply system;
- for reducing the energy demand or for changing the structure of the demand.

The model estimates the effects of the various strategies. It calculates the economical and environmental consequences and can test the strategies under various possible economic developments. On the basis of that information one can compare the various strategies and select those which have consequences that can be accepted by the whole system.

2. Work Programme

(a) Data analysis;

(b) Model modification;

(c) Data preparation;

(d) Analysis of the functional relations of the corresponding system values;

(e) Construction of the various modules of the model;

                                                             ../..

Total cost:
   DM.   475,000
   u.a.   129,781
	E.C. Contribution: 61%
	   DM. 292,000
	   u.a. 79,781

COMMISSION OF THE EUROPEAN COMMUNITIES ENERGY R & D PROGRAMME Objective: Systems analysis: development of models	Project or Sector:

2. Work Programme

(f) Validation of the modules of the model;

(g) Combining the different modules to the complete model;

(h) Validation of complete model.

3. Status

For further development and modification, the model has been
broken down into individual model sectors in accordance with
its modular structure. This makes it easier to share out
the work among the members of the group and facilitates co-
ordination. All model sectors are currently in the validation
stage, i.e., the values calculated have to be compared with the
statistical values in the data base.

COMMISSION OF THE EUROPEAN COMMUNITIES ENERGY R & D PROGRAMME Objective: Systems analysis: development of models	Project or Sector:

Title:
To adapt to the Netherlands case the model developed at the K.F.A. in Jülich for the European Communities.

Duration: 9 months Period: 1/10/76 - 30/6/77	Contract No: 049-76 EMN Project No:

Contractor: Centrale Organisatic voor Toegepast Natuurwetenschappelijk Onderzock (T.N.O.)

Address: Juliana van Stolberglaan, 148 Den Haag, Netherlands

Head of Project: Dr. Ir. A.J. Bogers

Description of research work

I. Objectives (aims)

(a) Qualitative evaluation of actual model capabilities;

(b) Data collection "Energy Supply Sector" of the model";

(c) Test runs and quantitative evaluation of "Energy Supply Sector".

2. Work Programme

(a) Qualitative evaluation : 1/12/1976 - 1/2/1977
(b) Data collection : 1/10/1976 - 1/3/1977
(c) Test runs and evaluation : 1/3/1976 - 10/6/1977

3. Status

(a) Paper available;"An assessment of the KFA-Jülich Energy Model" (Brascamp, Timman, Tweehuysen);

(b) Complete;

(c) Underway.

Total cost:
  Fl. 72,400
  u.a. 20,000
	E.C. Contribution: 100%
	  Fl. 72,400
	  u.a. 20,000

COMMISSION OF THE EUROPEAN COMMUNITIES ENERGY R & D PROGRAMME Objective: Systems analysis: development of models	Project or Sector:

Title:

    Demonstration of the application of the world petroleum
and natural gas model to EEC type problems.

Duration: 4 months Period: 1/9/76 - 31/12/76	Contract No: 050-76 EMUK Project No:

Contractor: Energy Models Limited

Address: Queen Mary College of the University of London
              Mile End Road, London E1 4NS, United Kingdom

Head of Project: Mr. R.J. Deam

Description of research work

  I. Objectives (aims)

  To demonstrate the use of the QMC World Petroleum and gas model in
the following applications:-

  (a) investigating the level, type, timing and location of future
     refining, shipbuilding and other energy industry investments;

  (b) estimating economic crude oil and gas production rates, supply
     patterns and price structures;

  (c) investigating the global and national implications of international,
     EEC or unilateral decisions on environmental and conservational
     policies relating to the energy industry;

  (d) investigating the effects of new technologies in the energy
     industries.

  2. Work Programme

  A series of demonstration runs have been carried out relating to 1984
to show the logistic and economic effects of certain environmental or
conservational policy decisions by EEC or other nations:-

                                        ../..

Total cost:
  £   12,500
  u.a. 30,000
	E. C. Contribution: 100%
	  £ 12,500
	  u.a. 30,000

COMMISSION OF THE EUROPEAN COMMUNITIES ENERGY R & D PROGRAMME Objective:　　Systems analysis: 　　　　　　　development of models	Project or Sector:

2. Work Programme cont/..

(a) appreciably reducing the sulphur level of distillate and
     residual fuel oil used in the European Community;

(b) banning the use of nuclear power generation in the Community;

(c) the USA prohibiting the export of North Slope Alaskan crude;

(d) the oil producers changing the price of crude oil.

3. Status

All the runs in the existing programme have been completed and the
report prepared.

An updated model representing 1985 is currently being prepared and
will be available shortly.

COMMISSION OF THE EUROPEAN COMMUNITIES ENERGY R & D PROGRAMME Objective: Systems analysis: development of models	Project or Sector:

Title:

Use of the EXPLOR-output model to implement the demand link to the CEC energy flow models.

Duration: 9 months Period: 1/6/76 - 28/2/77	Contract No: 054-76 EMD Project No:

Contractor: Battelle Institut

Address: Battelle-Institut e.V., Am Römerhof 35, 6000 Frankfurt/ Main 90, Germany

Head of Project: Dr. H. Mischke

Description of research work

I. Objectives (aims)

The CEC has agreed upon a design of a European energy model, which is called the EC Energy Model. This model consists of different modules, which are models of their own. Phase I, Level II of EC's Energy Modelling was defined to develop the methodology of the EC Energy Model and to establish the model for the test country France. Battelle's contribution is to adapt and update its input-output model EXPLOR and to develop and energy demand model in order to project the transition from monetary energy demand into physical energy demand by sector and energy product. The physical energy demand by energy product will be used by an energy flow model. The concern of Battelle, therefore, is to project the sectorial economic demand and to establish a link to the physical energy flow.

2. Work Programme

C1. Data research in order to establish the transition matrix between the I-O-Model and the LP for the country of France (mid May - end of July 1976).

C2. Updating EXPLOR data base for France and development of basic software for continous updating of EXPLOR (mid July 76 - end January 77).

C3. Development of an Energy Demand Model in order to project the transition matrix (July 76 - mid February 77).

../..

Total cost: DM. 200.000 u.a. 54.645	E.C. Contribution: 100% DM. 200.000 u.a. 54.645

COMMISSION OF THE EUROPEAN COMMUNITIES ENERGY R & D PROGRAMME Objective: Systems analysis: development of models	Project or Sector:

2. Work Programme cont/..

<u>C4</u>. Test of EXPLOR and the Energy Demand Model (February 77 - mid April 77).

3. Status

The most ambitious research done in the energy field combines Input-Output-Models (monetary units) with Energy-Flow-Models (physical units).   The ERDA project tries to link the Hoffman Energy-Flow-Model with the Hudson-Jorgenson Input-Output-Model.   The result of EC's approach should be similar to the ERDA project.

So far, Battelle has delivered a draft (end of January 1977) dealing mainly with the above-mentioned tasks C1 and C2.   The final report will be given at the end of April, including the results of all tasks of Phase I.

COMMISSION OF THE EUROPEAN COMMUNITIES ENERGY R & D PROGRAMME Objective: Systems Analysis: development of models	Project or Sector:

Title: Study on energy products costs.

Duration: 12 months Period: 1/10/76 - 30/9/77	Contract No: 085-76 EMI Project No:

Contractor: Applicazioni e Ricerche Scientifiche (A.R.S.) Address: Viale Maino, 35, Milano, Italy Head of Project: M. S. Albertoni

Description of research work

   I. Objectives (aims)

   (a) Determination of rational joint costs of energetic products such as refinery, electricity and steam, according to available processes and demand area;

   (b) Energy demand forecast on the transport field on the basis of the past by regression analysis.

   2. Work Programme

   (a) Application of the joint costs theory to the refinery technical "scenario";

   (b) Application of the cross impact analysis to the transport energy demand for Italy and England.

   3. Status

Work statement of item (a) is almost completed, while the item (b) is developed to the level of 30% of the total amount of the work to be done.

Total cost:
      Lit.   30,000,000
      u.a.   48,000
	E.C. Contribution: 100%
	      Lit. 30,000,000
	      u.a. 48,000

COMMISSION OF THE EUROPEAN COMMUNITIES ENERGY R & D PROGRAMME Objective: Systems Analysis: Development of Models	Project or Sector:

Title:

Reference Energy System Data Base and its Integration with the Energy Flow Optimization Model (Phase II).

Duration: 12 months Period: 1/1/77 - 31/12/77	Contract No: 210-77 EMB Project No:

Contractor: Systems Europe (SE)

Address: Avenue de Broqueville, 194, 1200 Brussels, Belgium

Head of Project: E. Jamoulle

Description of research work

I. Objectives (aims)

Scope of Phase II: Guidelines

Phase II is essentially "multinational oriented".   Its objective is to broaden the scope of single country models along two guidelines.

A first guideline is "horizontal extensions", i.e., to apply to other EEC member countries what was developed and applied to France as pilot country during Phase I (1976) namely for what refers to S.E.'s contribution:

- expansion of the Energy Model Data Base and the Reference Energy System to eight more countries;

- extension of the Simulation Model operating on the Data Base.

A second guideline is "vertical expansion", i.e., to develop a multi-national optimization model operating on a Multinational Reference Energy System with identification of both individual country's components and intercountry relations.

2. Work Programme

Group of tasks general description and organization:

1st Group Data Base Expansion and Maintenance

      T10 — Common Body   ../..

Total cost:
   FB.   6,000,000
   u.a.   120,000
	E.C. Contribution: 100%
	   FB. 6,000,000
	   u.a. 120,000

COMMISSION OF THE EUROPEAN COMMUNITIES ENERGY R & D PROGRAMME Objective: Systems Analysis: Development of Models	Project or Sector:

2. Work Programme cont/..

    T11 — Assist. in Data Collection: The Netherlands

    T12 — Assist. in Data Collection: Germany

    T13 — Assist. in Data Collection: Italy

    T14 — Assist. in Data Collection: United Kingdom

    T15 — Assist. in Data Collection: Ireland

    T16 — Execution of Data Collection Belgium and Luxembourg

    T19 — Assist in DB Maintenance

2nd Group: Software Extension and Improvements:

    T21 — DAMOCLES: Direct Access Management Organization for Computerized Limited Energy Systems

    T22 — SIMUL 2

3rd Group: Multinational Optimization Model (ORESTE)

    T31 — MERES: Multiregional European Reference Energy System

    T32 — MEEDB

    T33 — Single Country L.P.

    T34 — Multinational L.P. (ORESTE) (Optimization of Reference Energy Systems Transfers and Exchanges)

4th Group: Project Management and Organization

    T41 — Monitoring

    T42 — Reporting

3. Status

Ongoing.

Register of contractors

KEY TO CONTRACT NUMBERS

Each contract is identified by a sequence of figures and letters, the key to which is as follows.

First three figures belong to a numerical order system attributing a number to each contract as it arises.   The following two figures indicate the year in which the contract was concluded.

There follows a series of letters.   All letter series begin with the letter E which simply denotes that the contract is under the general heading of the Energy Research and Development Programme.

The second letter (there are five variations) denotes the objective ("sub-programme") of the Energy R & D programme:

    E....   Energy Conservation
    H....   Production and Utilization of Hydrogen
    S....   Solar Energy
    G....   Geothermal Energy
    M....   System Analysis: Development of Models

Following the second letter there can be either one, two or three letters denoting the country in which the contract will be carried out.

    B....   Belgium
    D....   Germany
    DK...   Denmark
    EIR..   Ireland
    F....   France
    I....   Italy
    N....   Netherlands
    UK...   United Kingdom

The bracketed letter at the end of each Contract No. refers to the project or sector within one of the previously mentioned and listed five objectives (E, H, S, G, or M.).

E - Energy Conservation:

  (a) improved insulation of buildings;

  (b) use of heat pumps;

  (c) urban transport;

  (d) residual heat recovery;

  (e) materials recycling;

  (f) production of energy from waste;

  (g) evaluation of the specific energy consumption of
       equipment, processes and techniques;

  (h) development of methods for storage of secondary energy.

H - Production and Utilization of Hydrogen:

  (A) Thermochemical production of hydrogen;

  (B) Electrolytic production of hydrogen;

  (C) Utilization of hydrogen.

S - Solar Energy:

  (A) Solar heat collectors and their application to dwellings;

  (B) Self-contained generating sets for the production of
       mechanical and/or electrical power;

  (C) Photovoltaic conversion;

  (D) Photochemical, photoelectrochemical and photobiological
       processes;

  (E) Photosynthetic production of organic matter;

  (F) Data network relating to solar radiation.

G - Geothermal Energy:

  (A) Acquisition and collation of existing and new geothermal
       data;

  (B) Improvement of methods of exploration;

  (C) Sources of hot water (low enthalpy);

  (D) Steam sources (high enthalpy) and hot rocks;

  (E) Training of specialists.

M - Systems Analysis - Development of Models.

B E L G I U M

Contractors	Contract No.	
Advanced Technology Research and Application Company (ATRAC)	117-76	ESB(A)
Centre d'Etudes de l'Energie Nucléaire (CEN)	060-76	EHB(B)
Centre de Recherches Scientifiques et Techniques de l'Industrie des Fabrications Métalliques (CRIF)	202-77	EEB(e)
Faculté Polytechnique de Mons	116-76	ESB(A)
Katholieke Universiteit Leuven (K.U.L.)	141-76	ESB(A)
	152-76	ESB(C)
Rijksuniversiteit Gent	154-76	ESB(C)
Systems Europe	010-76	EMB
	210-77	EMB
Université Catholique de Louvain U.C.L.	011-76	EMB
	174-76	EEB(g)
Universitaire Instelling Antwerpen	119-76	EGB(B)
	219-77	EGB(B)
Université Libre de Bruxelles (U.L.B.)	012-76	EMB
	028-76	ESB(D)

D E N M A R K

Contractors	Contract No.
Aarhus Universitet	102-76 EGDK(A)
Danmarks Tekniske Højskole	140-76 ESDK(A)
European Heat Pump Consultors	142-76 EEDK(b)
International Solar Power Co. Ltd.	137-76 ESDK(A)
Jordbrugsteknisk Institut	051-76 ESDK(E)
United Breweries Ltd.	029-76 ESDK(D)

F R A N C E

Contractors	Contract No.	
Amis de l'Institut Economique et Juridique de l'Energie (A.I.E.J.E.)	036-76	EMF
Association pour la Recherche et le Développement des Méthodes et Processus Industriels (ARMINES)	099-76	EGF(C)
Bureau de Recherches Géologiques et Minières (B.R.G.M.)	078-76	EGF(A)
	079-76	EGF(B)
	081-76	EGF(D)
	092-76	EGF(C)
	093-76	EGF(C)
	170-76	EGF(A)
	178-77	EGF(B)
Centre National d'Etudes Spatiales (CNES)	193-77	ESF(C)[1]
Centre National pour la Recherche Scientifique (C.N.R.S.)	013-76	ESF(D)
	080-76	EGF(B)
	082-76	EGF(B)
	095-76	EGF(A)
	100-76	EGF(D)
	101-76	EGF(B)
	103-76	EGF(B)
	147-76	EGF(B)
	172-76	ESF(B)
	180-77	EGF(B)
	206-76	ESF(C)[2]
	220-77	EGF(B)
Centre de Recherche de la Compagnie Générale d'Electricité (Laboratoires de Marcoussis)	062-76	EHF(B)
	179-77	EEF(h)
Cerchar	173-76	EEF(d)
	199-77	EEF(f)
Comité Scientifique et Technique de l'Industrie du Chauffage, de la Ventilation et du Conditionnement d'Air (COSTIC)	171-76	ESF(A)[3]

---

[1] with IRCHA

[2] with ONERA and Université Paul Sabatier

[3] with Présente Roulier and CEA

Contractors	Contract No.
Commissariat à l'Energie Atomique (CEA)	014-76　ESF(D)
	040-76　EHF(B)
	041-76　EHF(C)
	066-76　EHF(A)
	070-76　EHF(C)(1)
	118-76　ESF(A)
	125-76　ESF(C)
	171-76　ESF(A)(2)
	201-77　EEF(a)
Compagnie Electro-Mécanique (C.E.M.)	063-76　EHF(B)
Ecole Nationale Supérieure de Chimie de Paris	159-76　ESF(c)(3)
Electricité de France (EDF)	077-76　EGF(c)(4)
	094-76　EGF(C)
Fondation Ed. de Rotschild pour le Développement de la Recherche Scientifique	015-76　ESF(D)
Gaz de France (GDF)	043-76　EHF(C)
	160-76　EHF(C)
Heurtey S.A.	200-77　EEF(f)
Ingénierie Française du Tout Electrique (INFRATEL)	077-76　EGF(c)(5)
Institut Francais du Pétrole (I.F.P.)	069-76　EHF(B)(6)
	070-76　EHF(C)(7)
	122-76　EEF(c)
	123-76　EEF(c)
	124-76　EEF(g)
	164-76　EGF(D)
Institut National de la Recherche Agronomique	053-76　ESF(E)

---

(1) with Institut Francais du Pétrole (I.F.P.)

(2) withPrésente Roulier and COSTIC

(3) with Université du Languedoc et Université du Haut-Rhin

(4) with INFRATEL

(5) with EDF

(6) with S.R.T.I.

(7) with CEA

Contractors	Contract No.
Institut National de Recherche Chimique Appliquée (IRCHA)	193-77 ESF(c)[1]
Laboratoires d'Electronique et de Physique Appliquée (LEP)	109-76 ESF(C) 110-76 ESF(C)
Office National d'Etudes et de Recherches Aérospatiales (ONERA)	206-76 ESF(c)[2]
Présente Roulier Engineering S.A.	171-76 ESF(A)[3]
Radiotechnique-Compelec (R.T.C.)	106-76 ESF(C) 107-76 ESF(C)
Société Bertin et Cie.	114-76 ESF(A) 169-76 EEF(d)
Société Creusot-Loire S.A.	064-76 EHF(C)
Société de Recherches Techniques et industrielles (SRTI)	069-76 EHF(B)[4]
Société Nationale Elf Aquitaine (SNEA)	115-76 ESF(A)
Société Rhône Poulenc Industries	168-76 ESF(A)
Université du Haut-Rhin (Mulhouse)	159-76 ESF(C)[5]
Université du Languedoc (Montpellier II)	159-76 ESF(C)[6]
Université de Metz	065-76 EHF(C)
Université Louis Pasteur	076-76 EGF(B)
Université Paul Sabatier (Toulouse)	206-76 ESF(C)[7]

---

[1] with CNES
[2] with CNRS and Université Paul Sabatier
[3] with COSTIC and CEA
[4] with Institut Français du Pétrole
[5] with l'Université du Languedoc and Ec. Nat. Sup. de Chimie, Paris
[6] with l'Université du Haut-Rhin and Ec. Nat. Sup. de Chimie, Paris
[7] with CNRS and ONERA

G E R M A N Y

Contractors	Contract No.
Battelle Institut e.V.	054-76 EMD
	132-76 ESD(C)
	189-77 ESD(C)
	214-77 EED(b)
Bayerische Landesanstalt für Landtechnik	052-76 ESD(E')
Brown Boveri & Cie A.G.	045-76 EHD(B)
	129-76 ESD(A)
Deutsche Forschungs-und Versuchsanstalt für Luft-und Raumfahrt e.V. (DFVLR)	056-76 EHD(A)
	057-76 EHD(C)
Dornier System GmbH	046-76 EHD(C)
Forschungs-und Entwicklungslabor Kleinwächter	104-76 ESD(C)
Fraunhofer Gesellschaft zur Förderung der Angewandten Forschung e.V.	131-76 ESD(A)
Fritz Haber Institut der Max Planck Gesellschaft	025-76 ESD(D)
Gesellschaft für Praktische Energikunde (G.F.P.E.)	190-77 EED(g)
Kernforschungsanlage Jülich GmbH	048-76 EMD
	162-76 EHD(A)
Krebskosmo GmbH	058-76 EHD(B)
Maschinenfabrik Augsburg Nürnberg A.G. (M.A.N.)	211-77 EED(b)
Messerschmitt-Bölkow-Blohm GmbH (M.B.B.)	032-76 ESD(B)[1]
	130-76 ESD(A)
Niedersächsisches Landesamt für Bodenforschung	071-76 EGD(A)
	177-77 EGD(B)
Rheinisch-Westfälische Technische Hochschule Aachen	185-77 EHD(A)
Ruhr Universität Bochum	021-76 ESD(D)
	098-76 EGD(A)
Stadt Urach	176-77 EGD(A)
Technische Hochschule Darmstadt	047-76 EHD(B)

---

[1] with ANSALDO (Italy)

Contractors	Contract No.
Technische Universität Berlin	023-76 ESD(D)
	024-76 ESD(D)
Technische Universität Braunschweig	127-76 EGD(B)
Universität Göttingen	126-76 EGD(B)
Universität Karlsruhe	075-76 EGD(B)
Universität Tübingen	128-76 EGD(B)
Universität Würzburg	022-76 ESD(D)
Vereinigte Elektrizitätswerke Westfallen A.G. (V.E.M.)	181-77 EED(b)

I R E L A N D

Contractors	Contract No.
An Foras Taluntais	055-76 ESEIR(E)
Institute for Industrial Research and Standards	134-76 ESEIR(A)
University College Cork	059-76 EHEIR(A)
	163-76 ESEIR(C)
University College Dublin	143-76 EEEIR(o)

I T A L Y

Contractors	Contract No.	
Azienda Generale Italiana Petroli S.p.A. (AGIP)	089—76	EGI(D)
Analysis and Development of Energy Systems S.R.L. (ADES)	038—76	EHI(A)
Ansaldo	032—76	ESI(B)[1]
Applicazioni e Ricerche Scientifiche (A.R.S.)	085—76	EMI
Consiglio Nazionale Delle Ricerche (C.N.R.)	087—76	EGI(A)
	088—76	EGI(A)
	112—76	EGI(B)
	195—76	ESI(C)
Centro Informazione Studi ed Esperienze (CISE)	194—76	ESI(C)
Ente Nazionale per l'Energia Elettrica (ENEL)	086—76	EGI(C)
	166—76	EGI(D)
	167—76	EGI(D)
Fiat	105—76	ESI(A)
	196—76	EEI(d)
Gemmindustria SNC	198—77	EEI(d)
Montedison	186—76	ESI(C)
Nuovo Pignone S.p.A.	121—76	ESI(A)
Oronzio de Nora	067—76	EHI(B)
Osservatorio Vesuviano	165—76	EGI(A)
Politecnico di Milano	148—76	EHI(B)
Università Degli Studi di Bari	091—76	EGI(B)
Università di Bologna	031—76	ESI(D)
Università di Milano	030—76	ESI(D)
	039—76	EHI(B)
Università di Modena	150—76	ESI(C)
Università di Napoli	090—76	EGI(A)
	203—77	EGI(D)
Zanussi S.p.A.	149—76	ESI(A)

---

[1] with M.B.B. (Germany)

N E T H E R L A N D S

Contractors	Contract No.	
Centrale Organisatie voor Toegepast Natuur-wetenschappelijk Onderzoek T.N.O.	049-76	EMN
	061-76	EHN(B)
	073-76	EGN(A)
	133-76	ESN(A)
	144-76	EEN(a)
	231-77	EEN(b)
Fondation for Fundamental Research on Matter (Stichting F.O.M.)	156-76	ESN(C)
Katholieke Universiteit Nijmegen	111-76	ESN(C)
Netherlands Institut voor Zuivelonderzoek	215-76	EEN(g)
Rijksuniversiteit Utrecht	155-76	ESN(C)
Rijksuniversiteit te Leiden	027-76	ESN(D)
Stichtingen Bouwcentrum en Ratiobouw	138-76	ESN(A)
Technische Hogeschool te Eindhoven	182-77	EHN(B)
Universiteit van Amsterdam	026-76	ESN(D)

U N I T E D   K I N G D O M

Contractors	Contract No.
Building Services Research and Information Association (B. S. R. I. A.)	187-77 EEUK(g)
Central Electricity Generating Board (C. E. G. B.)	218-77 EGUK(D)
City University London	161-76 EHUK(B)
Energy Models Ltd.	050-76 EMUK
Ferranti Ltd.	146-76 ESUK(C)
General Technology System Ltd. (G. T. S.)	044-76 ESUK(B)
	205-77 ESUK(B)
Imperial College of Science and Technology	016-76 ESUK(D)
	097-76 EGUK(A)
International Research and Development Co. Ltd. (IRD)	175-77 EEUK(b)
	184-77 EEUK(d)
	191-76 ESUK(C)
John Laing Research and Development Ltd.	139-76 ESUK(A)[1]
Loughborough University of Technology	212-77 EEUK(a)
Members of the Royal Institution of Great Britain	017-76 ESUK(D)
Milton Keynes Development Corporation	139-76 ESUK(A)[2]
National Coal Board	068-76 EHUK(C)
	188-77 EEUK(f)
Natural Environment Research Council	074-76 EGUK(A)
	084-76 EGUK(A)
G.V. Planer Limited	145-76 ESUK(C)
Plessey Company Ltd.	113-76 ESUK(C)
	120-76 ESUK(C)
Polytechnic of Central London	139-76 ESUK(A)[3]

---

[1] with Polytechnic of Central London and Milton Keynes Development Corporation

[2] with Polytechnic of Central London and John Laing R & D Ltd.

[3] with Milton Keynes Development Corporation and John Laing R & D Ltd.

XII/697/77-E

Contractors	Contract No.
University of Bristol	018-76 ESUK(D)
University College Cardiff	136-76 ESUK(A)
University College London	019-76 ESUK(D)
University of Leeds	135-76 ESUK(A)
University of London King's College	020-76 ESUK(D)
University of Oxford	096-76 ECUK(A)
University of Reading	192-77 ESUK(E)

Annexes

I The council's programme decision

INDIRECT ACTION

ENERGY RESEARCH AND DEVELOPMENT PROGRAMME

The aims of the programme are as follows:

1. ENERGY CONSERVATION

    A maximum of 11.380 million units of account and a staff of six
    shall be assigned to this objective.

    This programme covers the following sectors:

    — improved insulation of buildings,

    — use of heat pumps,

    — urban transport,

    — residual heat recovery,

    — materials recycling,

    — production of energy from waste,

    — evaluation of the specific energy consumption of equipment,
       processes and techniques,

    — development of methods for storage of secondary energy.

    This work shall be carried out under contract.

2. PRODUCTION AND UTILIZATION OF HYDROGEN

    A maximum of 13.240 million units of account and a staff of four
    shall be assigned to this objective.

    This programme comprises the following projects:

    Project A: Thermochemical production of hydrogen

    Action: 1. Research into chemical and electrochemical reaction
                    cycles of high potential efficiency in the
                    conversion of heat energy into hydrogen energy;

                2. Practical experiments on promising cycles.

    Project B: Electrolytic production of hydrogen

    Action: 1. Improvement of existing electrolytic production
                    technology;

                2. Study of the viability and economics of high-
                    temperature and high-pressure electrolysis.

Project C: Utilization of hydrogen

Action: 1. Analysis of the potential applicability of hydrogen
               and of synthetic hydrogen-based fuels;

            2. Development of safety specifications for the handling
               of hydrogen;

            3. Improvement of the small-scale storage of hydrogen.

This work shall be carried out under contract.

3. SOLAR ENERGY

    A maximum of 17.500 million units of account and a staff of six
    shall be assigned to this objective.

    This programme comprises the following projects:

    Project A: Solar heat collectors and their application to dwellings

    Action: 1. Low-temperature use of solar energy for heating and
                   cooling buildings;

                2. Study of plane surface collectors;

                3. Pilot applications to dwellings for domestic use.

    Project B: Self-contained generating sets for the production of
               mechanical and/or electrical power

    Action: 1. The use in medium and high temperature areas of
                   solar heat to produce mechanical and/or electrical
                   power;

                2. Improvement of low-power groups (1 + 10 kw);

                3. Pilot installation of 1 MWe.

    Project C: Photovoltaic conversion

    Action: 1. Development of alternative cells and improvement of
                   existing cells;

                2. Feasibility study on new concepts;

                3. New methods of preparing semiconductor materials;

                4. Silicon thin film;

                5. Automation of panel production.

Project D: Photochemical, photoelectrochemical and photobiological
             processes

Action: 1. Basic studies on photochemical, photoelectrochemical
                 and photobiological systems.

Project E: Photosynthetic production of organic matter

Action: 1. Choice and development of the most suitable energy
                 crops for the different regions of Europe.

Project F: Data network relating to solar radiation

Action: 1. Collection, standardization and distribution of
                 comprehensive data on number of hours of sunshine
                 throughout the Community;

             2. Definition of the implications of the large-scale
                 use of solar energy.

This work shall be carried out under contract.

4.   GEOTHERMAL ENERGY

A maximum of 13.000 million units of account and a staff of four
shall be assigned to this objective.

This programme comprises the following projects:

Project A: Acquisition and collation of existing and new geothermal
             data

Action: 1. Collation of existing geothermal data;

             2. Acquisition and collation of new additional
                 geothermal data.

Project B: Improvement of methods of exploration

Action: 1. Improvement and/or adaptation of existing prospecting
                 methods to specific geothermal requirements and
                 development of new methods of prospecting and
                 exploration.

Project C: Sources of hot water (low enthalpy)

Action: 1. Compilation of geothermal models in regions
                 concerned;

        2. Full-scale experimental verification of theoretical models (operation);

        3. Utilization tests on sources of hot water for district and agricultural heating.

Project D: Steam sources (high enthalpy) and hot rocks

Action: 1. Construction of geothermal models in the areas concerned;

        2. Improvement of measuring and drilling techniques for experimental work at high temperatures;

        3. Stimulation of hot rocks to increase their permeability and extraction of heat from hot rocks.

Project E: Training of specialists

Action: 1. Training courses and detachments.

This work shall be carried out under contract.

5. SYSTEMS ANALYSIS: DEVELOPMENT OF MODELS

A maximum of 3.880 million units of account and a staff of seven shall be assigned to this objective.

This programme comprises the following action:

Action: 1. Static models (short term);

        2. Dynamic sector models (medium/long term).

This work shall be carried out under contract.

II Project leaders (expert rapporteurs)

ENERGY CONSERVATION

Prof. Dr. Ing. L. GÖTTSCHING
Technische Hochschule Darmstadt
Germany

Ir. J. A. KNOBBOUT
Centrale Organisatie voor Toegepast
Natuurwetenschappelijk Onderzoek (TNO)
Netherlands

PRODUCTION AND UTILIZATION
OF HYDROGEN

Project A: Ing. G. BEGHI (C.E.C.)
Joint Research Centre Ispra
Italy

Project B: Prof. Dr. H. WENDT
Technische Hochschule Darmstadt
Germany

Project C: M. G. DONAT
Gaz de France
France

SOLAR ENERGY

Project A: M. ARANOVITCH (C.E.G.)
Joint Research Centre Ispra
Italy

Project B: Dr. GRETZ (C.E.C.)
Joint Research Centre Ispra
Italy

Project C: Prof. Dr. Ir. R. VAN OVERSTRAETEN
                  Katholieke Universiteit Leuven
                  Belgium

Project D: Prof. Dr. D. HALL
                  University of London, King's College
                  United Kingdom

Project E: Dr. E. LALOR
                  National Science Council
                  Ireland

GEOTHERMAL ENERGY

Project A: Dr. R. HAENEL
                  Niedersächsisches Landesamt
                  für Badenforschung
                  Germany

Project B: M. F. MUNCK
                  Service Géologique National (S.G.N.)
                  France

Project C: M. A. GRINGARTEN
                  Service Géologique National (S.G.N.)
                  France

Project D: Dr. R. CATALDI
                  Ente Nazionale per l'Energia
                  Elettrica (ENEL)
                  Italy

III Members of the advisory committees on programme management (ACPM)

President : Mr. G. PRESTON
Department of Energy
London

BELGIUM

M. J. PONCIN
Programmation de la Politique Scientifique
Bruxelles

M. MARKEY
Ministère des Affaires Economiques
Administration de l'Energie
Bruxelles

M. G. F. A. de VOGELAERE
Ministère des Affaires Economiques
Administration de l'Energie
Bruxelles

M. J. PATIGNY
Université Catholique de Louvain
Louvain

M. UYTTENBROECK
Centre Scientifique et Technique
de la Construction
Bruxelles

DENMARK

Hr. B. QVALE
Danmarks tekniske Højskole
Lyngby

Hr. K. DAVIDSEN
Københavns Belvsningsvaesen
København

Hr. S. R. JACOBSEN
Dansk Kedelforening
Hellerup

FRANCE

M. R. DUMON
Société Heurtey
Paris

M. P. LEPRINCE
Institut Français du Pétrole
Rueil-Malmaison

M. C. PALVADEAU
Agence pour les économies d'énergie
Paris

GERMANY

Herr. H. KLEIN
Bundesministerium für Forschung
und Technologie (BMFT)
Bonn

Herr. U. PLANTIKOW
Kernforschungsanlage Jülich (KFA)
Julich

IRELAND

Mr. T. J. QUINN
Institute for Industrial Research
and Standards
Dublin

Mr. P. T. PIGOTT
An Foras Forbatha Teo
Dublin

Mr. J. G. DUGGAN
National Science Council
Dublin

ITALY

Sig. A. ROSSI
Centro Ricerche FIAT
Torino

Sig. M. RAVEGGI
Centro Ricerche FIAT
Torino

Sig. V. MAZZAGLIA
Consiglio nazionale delle ricerche
Roma

NETHERLANDS

Hr. J. A. KNOBBOUT
Centrum voor Energievraagstukken TNO
Apeldoorn

Hr. G. KOL
Ministerie van Economische Zaken
Den Haag

Hr. K. WASSENAAR
N.V. KEMA
Arnhem

UNITED KINGDOM

Mr. P. G. O'NEILL
Department of the Environment
London

Mr. J. A. CATTERALL
Department of the Environment
London

Mr. J. BUTTERWORTH
Atomic Energy Research Establishment (AERE)
Harwell

ADVISORY COMMITTEE ON PROGRAMME MANAGEMENT (ACPM)

"PRODUCTION AND UTILIZATION OF HYDROGEN"

President : Mr. G. BERGES
C. E. A.
Paris

BELGIUM

M. DELHASSE
Services de Programmation de la
Politique Scientifique
Bruxelles

M. VAN DAMME
Ministère des Affaires Economiques
Bruxelles

M. FONTENOY
Services de Programmation de la
Politique Scientifique
Bruxelles

M. L. BAETSLE
C.E.N. - S.C.K. Laboratories
Mol - Donk

DENMARK

Hr. H. BOHLBRO
Haldor Topsoë A/S
Soeborg

Hr. L. T. MUUS
Aarhus Universitets
Aarhus

Hr. J. KOFOED
Danmarks Tekniske Højskole
Lyngby

FRANCE

Mr. P. COURVOISIER
C.E.N. Saclay
Gif-sur-Yvette

M. J. RASTOIN
C.E.N. Saclay
Gif-sur-Yvette

M. J. POTTIER
Gaz de France
Paris

M. P. GODIN
Electricité de France (EDF)
Chatou

GERMANY

Herr. F. FETTING
Technische Hochschule
Darmstadt

Herr. H. G. ROERTGEN
Rheinische Braunkohlenwerke
Aktiengesellschaft
Köln

Herr. J. JESSENBERGER
Bundesministerium für Forschung
und Technologie
Bonn

Herr. A. ZIEGLER
Bundesministerium für Forschung
und Technologie
Bonn

IRELAND

Mr. J. CUNNINGHAM
University College
Cork

Mr. J. GRAHAM
Department of Transport and Power
Dublin

ITALY   Sig. R. DANESI
                        C.N.E.N.
                        Roma

                        Sig. A. FACCHINI
                        AGIP Nucleare
                        Milano

                        Sig. G. C. SCIBONA
                        C.N.E.N.
                        Roma

LUXEMBOURG   M. Theisen
                        Nucleaire S.A.R.L.
                        Luxembourg

NETHERLANDS   Hr. M. E. A. HERMANS
                        KEMA
                        Arnhem

                        Hr. VAN DER PLAS
                        KEMA
                        Arnhem

                        Hr. J. QUAKERNAAT
                        TNO
                        Apeldoorn

                        Hr. J. A. A. KETELAAR
                        Akzo-Zout-Chemie
                        Hengelo

UNITED KINGDOM   Mr. J. K. DAWSON
                        Atomic Energy Research Establishment
                        Harwell

                        Mr. G. S. DEARNLEY
                        Department of Energy
                        London

ADVISORY COMMITTEE ON PROGRAMME MANAGEMENT (ACPM)

"SOLAR ENERGY"

President : Mr. R. CHABBAL
Centre National de la Recherche Scientifique
(C.N.R.S.)
Paris

BELGIUM   M. R. VANDAMME
                       Ministère des Affaires Economiques
                       Bruxelles

                       M. VAN VAERENBERGH
                       Service de la Programmation de la Politique
                       Scientifique
                       Bruxelles

DENMARK   Hr. V. KORSGAARD
                       Denmarks Tekniske Højskole
                       Lyngby

                       Hr. S. A. SVENDSEN
                       Denmarks Tekniske Højskole
                       Lyngby

                       Hr. O. LEISTIKO
                       Denmarks Tekniske Højskile
                       Lyngby

FRANCE   M. H. DURAND
                       Laboratoire d'Electronique et de Physique
                       Appliquée (L.E.P.)
                       Limeil-Brevannes

                       M. J. PHELINE
                       Délégation Générale à l'Energie
                       Paris

GERMANY

Herr. H. KLEIN
Bundesministerium für Forschung
und Technologie
Bonn

Herr. F. J. FRIEDRICH
Kernforschungsanlage Jülich
Julich

IRELAND

Mr. M. NEENAN
Agricultural Institute
Carlow

Mr. H. CLYNE
Institute for Industrial Research and
Development
Dublin

Mr. D. KEARNEY
National Science Council
Dublin

ITALY

Sig. G. BEER
Ansaldo S.p.A.
Genova

Sig. M. GHIO
Ministero del Tesoro R.G.S.
Roma

Sig. D. RAGUSA
E.N.I.
Roma

NETHERLANDS

Hr. C. J. HOOGENDOORN
Technische Hogeschool Delft
Delft

Hr. A. H. M. KIPPERMAN
Technische Hogeschool Eindhoven
Eindhoven

Hr. R. VAN DER WART
Energieonderzoek Centrum Nederland
Petten

UNITED KINGDOM   Mr. LONG
                        Atomic Energy Research Establishment (AERE)
                        Harwell

                        Mr. P. O'NEILL
                        Department of Environment
                        London

ADVISORY COMMITTEE ON PROGRAMME MANAGEMENT (ACPM)

"GEOTHERMAL ENERGY"

President : Sig. T. LEARDINI
Ente Nazionale per l'Energia
Elettrica (ENEL)
Roma

BELGIUM   M. DELMER
Ministère des Affaires Economique
Bruxelles

M. VAN VAERENBERGH
Service de Programmation de la Politique
Scientifique et Technique
Bruxelles

M. H. FONTENOY
Service de Programmation de la Politique
Scientifique et Technique
Bruxelles

M. LEGRAND
Ministère des Affaires Economiques
Bruxelles

M. de MAGNEE
Université Libre de Bruxelles
Bruxelles

M. BRYCH
Faculté Polytechnique de Mons
Mons

DENMARK   Hr. L. MADSEN
Danmarks Geologiske Undersoegelse
Koebenhavn

Hr. N. BALLING
Aarhus Universitet
Aarhus

FRANCE   M. P. SANGNIER
                          Délégation Générale à la Recherche
                          Scientifique et Technique (DGRST)
                          Paris

                          M. F. CHENEVIER
                          Ministère de l'Industrie
                          Paris

                          M. J. LAVIGNE
                          Bureau de Recherches Géologiques
                          et   uières
                          Orleans

GERMANY   Herr. J. JESSENBERGER
                          Bundesministerium für Forschung
                          und Technologie
                          Bonn

                          Herr. R. NEUMANN
                          Kernforschungsanlage Jülich
                          Jülich

IRELAND   Mr. C. R. ALDWELL
                          Geological Survey Office
                          Dublin

ITALY   Sig. E. BARBIER
                          Istituto Geotermico Internazionale
                          Pisa

                          Sig. C. CORVI
                          Ufficio Ministero Ricerca Scientifica
                          Roma

                          Sig. G. CHIERICI
                          AGIP
                          Milano

NETHERLANDS

Hr. R. VAN DER WART
Energieonderzoek Centrum Nederland
Petten

Hr. A. Johan BOOMER
Shell Internationale Petroleum Maatschappij
Den Haag

Hr. G. BROUWER
Ministerie van Economische Zaken
Haarlem

UNITED KINGDOM

Mr. J. D. GARNISH
Atomic Energy Research Establishment (AERE)
Harwell

Mr. R. A. ASHBEE
Central Electricity Generating Board (CEGB)
Leatherhead

Mr. W. BULLERWELL
Institute of Geological Sciences
London

ADVISORY COMMITTEE ON PROGRAMME MANAGEMENT (ACPM)

"SYSTEMS ANALYSIS : DEVELOPMENT OF MODELS"

President : Mr. P. McALISTER
National Science Council
Dublin

BELGIUM

Mme. F. THYS
Université Libre de Bruxelles
Bruxelles

M. Y. SMEERS
Université Catholique de Louvain
Heverlee

M. DELCROIX
Service de Programmation de la Politique
Scientifique
Bruxelles

M. H. B. PONCIN
Service de Programmation de la Politique
Scientifique
Bruxelles

DENMARK

Hr. B. ELBAEK
Atomenergikommissionens forsosgsanlaeg
risoe
Roskilde

Hr. N. MEYER
Danmarks Tekniske Højskole
Lyngby

Hr. A. HANSEN
Handelsministeriet
København

FRANCE

M. S. BINDEL
Délégation générale à la Recherche
Scientifique et Technique
Paris

M. D. BLAIN
Délégation générale à l'Energie
Paris

M. I. SACHS
Centre International de Recherche sur
l'Environnement et le Développement
Paris

M. I. THIRIET
Commissariat à l'Energie Atomique
Paris

GERMANY

Herr. T. BOHN
Kernforschungsanlage Jülich (KFA)
Jülich

Herr. W. ZUCKSCHWERDT
Bundesministerium für Forschung
und Technologie
Bonn

Herr. P. GONSCHIOR
Deutsche Entwicklungsgesellschaft
Köln

Herr. U. SCHOLER
Kernforschungsanlage Jülich (KFA)
Jülich

IRELAND

Mr. E. HENRY
Economic and Social Research Institute
Dublin

Mr. R. JOHNSTON
Trinity College Dublin
Dublin

ITALY

Sig. S. ALBERTONI
Societa ARS
Milano

Sig. M. CHIO
Ministero del tesoro
Roma

Sig. A. FEDRICHINI
Centre Ricerche FIAT
Torino

NETHERLANDS

Hr. A. J. BOGERS
Centrum voor energie T.N.O.
Apeldoorn

Hr. K. NATER
Energieonderzoek Centrum Nederland
Petten

Hr. A. A. J. SMEULDERS
Ministerie van Economische Zaken
Den Haag

Hr. M. J. STOFFERS
Centraal planbureau
Den Haag

UNITED KINGDOM

Mr. J. H. CHESSHIRE
University of Sussex
Sussex

Mr. BUTTERWORTH
Atomic Energy Research Establishment (AERE)
Harwell

Mr. F. W. BUTLER
Department of Energy
London

IV Members of the scientific and technical research committee (crest)

President : M. G. SCHUSTER
Director-General
Directorate-General "Research, Science and Education"
Commission of the European Communities

BELGIUM M. A. STENMANS
Service de Programmation de la Politique Scientifique
Bruxelles

M. WAUTREQUIN
Service de Programmation de la Politique Scientifique
Bruxelles

M. H. VAN HOUTTE
Représentation Permanente de la Belgique auprès des Communautés européennes
Bruxelles

DENMARK Hr. E. LYRTOFT PETERSEN
Udenrigsministeriet
København

Hr. P. A. KOCH
Forskningssekretariatet
København

Hr. P. BRANNER
Udenrigsministeriet
København

Hr. J. PEDERSEN
Forskningssekretariatet
København

FRANCE   M. GREGORY
                          Délégation Générale à la Recherche
                          Scientifique et Technique (DGRST)
                          Paris

                          M. F. RENOUARD
                          Représentation Permanente de la France
                          auprès des Communautés européennes
                          Bruxelles

                          M. Ph. PELTIER
                          Délégation Générale à la Recherche
                          Scientifique et Technique (DGRST)
                          Paris

GERMANY   Herr Dr. LEHR
                          Bundesministerium für Forschung und Technologie
                          Bonn

                          Herr J. BÖTTGER
                          Bundesministerium für Wirtschaft
                          Bonn

                          Herr E. LEYSER
                          Bundesministerium für Wirtschaft
                          Bonn

                          Herr G. VON KLITZING
                          Bundesministerium für Forschung und Technologie
                          Bonn

IRELAND   Mr. MANAHAN
                          Department of Industry and Commerce
                          Dublin

                          Mr. S. NIELSEN
                          National Science Council
                          Dublin

ITALY

Sig. P. BISOGNO
Consiglio Nazionale delle Ricerche (CNR)
Roma

Sig. S. FERRARI
Comitato Nazionale per l'Energia Nucleare
(CNEN)
Roma

Sig. G. CANDELARI
Ministro per la Ricerca Scientifica e
Tecnica (MRST)
Roma

Sig. M. LEVI
Ministro per la Ricerca Scientifica e
Tecnica (MRST)
Roma

LUXEMBOURG

M. H. HEYART
Ministère des Affaires Culturelles
Luxembourg

NETHERLANDS

Hr. C. H. STEFELS
Ministerie van Onderwijs en Wetenschappen
Den Haag

Hr. C. J. D. RIETHOF
Ministerie van Economische Zaken
Den Haag

Hr. J. D. de HAAN
Ministerie van Onderwijs en Wetenschappen
Den Haag

Hr. W. VISSER
Ministerie van Economische Zaken
Den Haag

Hr. J. H. W. FIETELAARS
Permanente Vertegenwoordiging van Nederland
bij de Europese Gemeenschappen
Brussel

UNITED KINGDOM   Ms. K. E. BOYES
                        Departments of Trade and Industry
                        London

                        Mr. H. M. G. STEVENS
                        Departments of Trade and Industry
                        London

COMMISSION   M. G. SCHUSTER
                        Président of the Committee
                        Bruxelles

                        M. VILLANI
                        Joint Research Centre
                        Bruxelles

                        M. J. LANNOY
                        Directorate-General "Scientific and Technical
                        Information and Information Management"
                        Luxembourg

                        M. H. von MOLTKE
                        Directorate-General "Internal Market and
                        Industrial Affairs"
                        Bruxelles

V Members of the crest subcommittee on energy research and development

President : M. P. DE MEESTER
Katolieke Universiteit Leuven
Belgium

BELGIUM

M. J. PONCIN
Service de Programmation de la Politique
Scientifique
Bruxelles

M. J. LIZIN
Administration de l'Industrie
Bruxelles

M. R. VAN DAMME
Administration de l'Energie
Bruxelles

M. P. MARKEY
Administration de l'Energie
Bruxelles

DENMARK

Hr. N. O. GRAM
Handelsministeriet
København

Hr. P. SNARE
International Energy Agency
København

FRANCE

M. S. BINDEL
Délégation Génerale à la Recherche
Scientifique et Technique (DGRST)
Paris

M. B. BAILLY du BOIS
Délégation Générale à la Recherche
Scientifique et Technique (DGRST)
Paris

M. FAUVE
Délégation Générale à l'Energie
Paris

M. J. PHELINE
Délégation Générale aux Energies Nouvelles
Paris

GERMANY

Herr SCHROETER
Bundesministerium für Forschung und
Technologie (BMFT)
Bonn

Herr ZIEGLER
Bundesministerium für Forschung und
Technologie (BMFT)
Bonn

Herr. H. V. STARKMUTH
Bundesministerium für Wirtschaft
Bonn

Herr J. ARNOLD
Ständige Vertrekungder BRD bei den E.G.
Bruxelles

IRELAND

Mr. P. McALISTER
National Science Council
Dublin

Mr. J. W. GRAHAM
Department of Transport and Power
Dublin

ITALY

Sig. C. CORVI
Ministero per la Ricerca Scientifica e
Tecnica (MRST)
Roma

Sig. F. BUSI
Consiglio Nazionale delle Ricerche (CNR)
Bologna

Sig. P. VENDITTI
Comitato Nazionale per l'Energia Nucleare
(CNEN)
Roma

Sig. F. REALE
Università di Napoli
Napoli

LUXEMBOURG   M. T. WEHENKEL
                          Représentation permanente du Luxembourg
                          auprès des Communautés européennes
                          Bruxelles

NETHERLANDS   Hr. R. F. de BRUINE
                          Ministerie van Economische Zaken
                          Den Haag

                          Hr. R. VAN DER WART
                          Energieondersoek Centrum Nederland (ECN)
                          Petten

UNITED KINGDOM   Mr. G. PRESTON
                          Department of Energy
                          London

                          Mr. R. SKIPPER
                          Department of Energy
                          London

                          Mr. T. MUIR
                          Office of the United Kingdom Permanent
                          Representative to the EC
                          Bruxelles

                          Mr. L. HARRIS
                          Office of the United Kingdom Permanent
                          Representative to the EC
                          Bruxelles

          Secretariat : M. M. FRANCINI
                        C. E. C.
                        Directorate-General "Research, Science
                        and Education"
                        Bruxelles

VI Information and patents

Section A - Information

A - 1   *The Commission may use for its own purposes the information*
         derived from the research programme defined in Annex I and the
         reports referred to in Article 3 of the General Terms and
         Conditions.

A - 2   *The Commission shall have the right to transmit in confidence the*
         reports referred to in Article 3 of the General Terms and
         Conditions, to Member States of the Community and to persons,
         bodies and undertakings engaged in the territory of a Member
         State of the Community in research or production justifying
         access to such reports.

A - 3   *The Commission shall have the right to publish without res-*
         triction the final report and the summary report referred to in
         Article 3 of the General Terms and Conditions, if the contractor
         has not, in a reasoned statement issued when such reports were
         being transmitted to the Commission, made publication thereof
         conditional on his approval. The contractor may refuse such
         approval only on condition that he does not publish himself.

A - 4   *The Commission shall state in every publication and communication*
         that the contractor is author of the reports unless the latter is
         opposed thereto.

A - 5   *The Commission shall also have the right to transmit the regular reports, final report and summary report referred to in Article 3 of the General Terms and Conditions to a non-member state or international organization under agreements and conventions concluded by the Community in accordance with the conditions laid down in Article 228 of the Treaty establishing the European Economic Community.*

     If such reports must be transmitted before publication the Commission shall inform the contractor thereof in writing. The latter may oppose transmission if he establishes that it would harm his industrial or commercial interests.

A - 6   *The contractor shall exchange on a reciprocal basis with other Commission contractors carrying out research related to that specified in Annex I, the information and reports referred to in paragraph A - 1. Such contractors will be designated to him by the Commission.*

     The procedures governing such exchange shall be the subject of an agreement between the Commission and the contractors concerned.

A - 7   *The contractor may use the information and reports referred to in paragraph A - 1 for his own purposes. He shall also have the right to transmit them to third parties or to publish them provided mention is made that they are derived from a contract concluded with the Community.*

     However, transmission to third parties or publication by the contractor of such information and reports prior to publication by the Commission must be approved by the Commission and shall be subject to conditions laid down by the latter.

A - 8   If the contractor passes on to the Commission information that
          may be patented, the parties shall refrain from any act which
          might prejudice the patentability of the invention until and not
          later than the filing of a first application for a patent.

Section B - Inventions

B - 1   Inventions, whether or not patentable, made or conceived when
          carrying out the programme specified in Annex I shall belong to
          the contractor if he so desires. Where the inventions are
          patentable he may apply for and obtain in his name the patents
          necessary for their protection, provided the Commission is
          informed thereof.

          In the three months following the transmission to the Commission
          of the final report referred to in Article 3 of the General Terms
          and Conditions the contracting parties shall draw up the list and
          descriptions of such inventions.

B - 2   If the contractor refrains from applying for the patents referred
          to in paragraph B - 1, the Commission may apply for and obtain
          the same on behalf of the Community. The contractor shall notify
          the Commission of his intentions in good time and in any event
          before the expiry date of the convention priorities, if any.

B - 3   The contractor undertakes to exploit or have the inventions and
          patents referred to in paragraph B - 1 exploited in conformity
          with the Community interest and to commence such exploitation not
          later than three years from the date of expiry of this contract.

B - 4   The Commission may propose to contractors who are carrying out or
          have carried out related research to conclude under the Community
          programme an agreement on the joint industrialization and
          marketing of the inventions, whether or not patentable, made or
          conceived when carrying out their contract with the Commission.

If such an agreement cannot be concluded these contractors may at
the request of the Commission obtain a licence in respect of the
inventions referred to in paragraph B - 1. This licence shall be
granted under conditions, in particular financial conditions,
determined taking into account, where appropriate, the contribution of the applicant to the research which led to the
invention.

B - 5   In respect of the inventions referred to in paragraph B - 1, the
Commission shall be entitled to a royalty-free licence for its
own purposes. This licence shall include the right to grant sub-
licences on the conditions specified in paragraph B - 6 hereafter
where the contractor fails without legitimate reason to fulfil
his obligation to exploit the said inventions or have them
exploited.

B - 6   The Commission may grant the sub-licences referred to in
paragraph B - 5 above to Member States and to persons and
undertakings which engage in the territory of a Member State in
research or production justifying the granting of a sub-licence
and which undertake to manufacture effectively in the Community.

The Commission shall advertise by all appropriate means offers
for the granting of non-exclusive sub-licences. If no applications are received pursuant to such offers the Commission shall
publish them in the Official Journal of the European Communities.
If within six months from the date of publication no application
has been submitted, the Commission may offer and grant exclusive
sub-licences for a maximum period of five years. It is understood, however, that the exclusive sub-licensee cannot oppose the
exploitation by the contractor of the inventions but that the
latter can no longer grant licences during the period of validity
of the exclusive sub-licence.

B - 7   *Before granting the sub-licences referred to in paragraph B - 5,*
         the Commission informs the contractor of its intention to grant
         the same and invites the contractor and the applicant for the
         sub-licence to conclude a licence contract between themselves. In
         case the contract has not been concluded within three months
         after such invitation, the Commission may grant the sub-licence
         at apporpiate conditions to be established, after consultation
         of the contractor, in mutual agreement between the Commission and
         the applicant for the sub-licence. The royalties paid by the
         sub-licensee are shared between the Community and the contractor,
         taking into account their respective financial contributions and
         the costs for industrial property rights, which might have been
         paid by them.

B - 8   *The contractor shall be entitled to a royalty-free, non-*
         exclusive, irrevocable licence in respect of the patents referred
         to in paragraph B - 2 above. If, after an advertising campaign
         identical to that provided for in paragraph B - 6, its offers of
         non-exclusive licences have not attracted applications, the
         Commission may convert this licence into an exclusive licence on
         terms to be agreed.

         If the contractor is not interested in an exclusive licence and
         the Commission succeeds eventually in granting one, such a
         licence may in no event be opposed to the contractor.

Section C - Basic information and patents

C - 1   *The contractor and the Commission undertake to furnish one*
         another with all the information which they have at their
         disposal and which would be necessary for the proper imple-
         mentation of the programme specified in Annex I or for the
         exploitation by themselves of the results obtained when carrying
         out this contract.

         If this information is confidential disclosure shall be on terms
         to be agreed.

C - 2   *The contracting parties undertake to grant on terms to be agreed non-exclusive licences in respect of the patents which they have at their disposal and which cover inventions made or conceived outside of the implementation of the programme specified in Annex I where the information referred to in paragraph A - 1 or the licences provided for in paragraphs B - 4, B - 5, B - 6 and B - 7 of this Annex cannot be exploited without infringing the said patents.*